The Golden Age of Homespun

This volume is published by Hill and Wang for
the New York State Historical Association

THE
Golden Age of Homespun

JARED VAN WAGENEN, JR.

ILLUSTRATIONS BY ERWIN H. AUSTIN

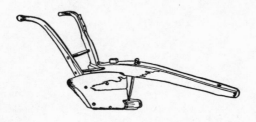

American Century Series
HILL AND WANG · NEW YORK
A division of Farrar, Straus and Giroux

Standard book number ISBN 0–8090–0066–0
Copyright 1953 by the New York State Historical Association
All rights reserved

FIRST AMERICAN CENTURY SERIES EDITION AUGUST 1963

Manufactured in the United States of America

10   11   12   13   14   15   16   17   18   19

FOR MAGDALENA, wife of my youth and of my sunset years

Foreword

THIS BOOK was written by one, who, over a long life, has risen to milk his cows, to plow his fields, to plant and garner with his own work-worn hands. His life has spanned the great arc of time which connects the last days of the oxen on our farms with these days of the modern combine. As a farmer he has always moved forward with the changes wrought in agricultural technology, but as a scholar he has sought to bring to light the moving story of life on the farms of our state in the days when hand skills were the first key to survival. Here is the harvest of his seeking, and it is a harvest such as no other has brought in.

It brings answers to a thousand passing questions that have drifted through our minds while reading other men's books or when we gazed, with little understanding, at some outworn utensil found in a cobwebby corner of loft or attic. You have seen neglected oxbows, but what do you know of their making or of the training of a yoke of oxen? Perhaps in your parlor there is a shoemaker's bench bought at a fancy price from an antique dealer, but what do you know of the rambling shoemakers who came to a farmhouse and stayed until each member of the family was newly shod with leather from the farm's cattle? Have you ever wondered about the processes by which our frontiersmen translated forest land into fields of wheat? What do you know about those two first crops of the pioneers, ashes and maple sugar? What do you know of log houses, of shingle making, bridges, and flax growing, of spinning and weaving cloth for a garment that was home-grown and homemade? Here is folk history, the accumulated memory of old men whom the author knew, but against whose memories he has held up the yardstick of the written record; the memories he has substantiated by a lifetime of research.

There is a double appropriateness to the fact that this volume is the first to be published under a Dixon Ryan Fox Memorial Fellowship granted by the New York State Historical Association. Dixon Ryan Fox, long the president of the Association, emphasized throughout his distinguished career as a historian the importance of understanding what went on in the everyday lives of the mass of men, and he would, I am confident, have applauded this book with all the enthusiasm that was one of his great assets.

In recent years, the Association has put its stress on social history, rather than on political and military history, particularly on the ways of life followed by those who lived simple, unnoticed lives. The Farmers' Museum, which the Association operates, seeks to tell three-dimensionally the story Jared van Wagenen has told in these pages. It is no accident that when Erwin Austin came to illustrate the book he found in the Museum source material for more than four-fifths of his pictures. One of the reasons for this is that since its inception a decade ago, Jared van Wagenen has been an active and creative force in the growth and development of the Farmers' Museum, of which he is a trustee.

Jared van Wagenen is a man all of one piece: the qualities he has admired in the people of the frontier have been the qualities he too has exemplified. Throughout the book and woven into its texture are his strong sense of moral values, his dry humor, his deep love for the world of his fathers which he has made his own, and love, too, for the land itself and those who till it with a steady hand.

Louis C. Jones, DIRECTOR
NEW YORK STATE HISTORICAL ASSOCIATION

Cooperstown, New York

Preface

I HAVE always deemed it a circumstance of distinct good fortune that I was farmborn. It happens that all my life of more than eighty-two years has been built around some 250 acres of tilled fields, hillside pasture, and pleasant woodland here in east-central New York. From these lands my people have made both a living and a life since great-grandfather set up his hearthstone here in the summer of 1800.

The September following my sixteenth birthday I went, or perhaps more correctly speaking was sent, to be a freshman in the embryonic Department of Agriculture in Cornell University. As a boy fresh from the fields and with stern, Puritanical standards, I was a vast remove from the glamorous personage sometimes visualized as a typical "college man." By some standards, my life was drab, but in my own thinking those four years were wonderfully good. One great opportunity they brought to me—the privilege of knowing in a fairly intimate way two very wonderful teachers and inspirers of youth. One was I. P. Roberts, head of the department, and the other was Liberty Hyde Bailey, who was to be Roberts' successor and dean. To them I owe more than I can tell, and after all the years I think of them with sentiments to which I can apply no colder term than affection.

I had gone to college with the distinct understanding that after graduation I was to return to the farm, and, now that so many years are past, I may say with truth that having thus put my hand to the plow, I have never looked back.

Three or four years after college, I became associated with that type of university extension known as the Farmers' Institute movement. This took the form of migratory pilgrimages about the state, with addresses to farmers when and wherever an audience could be gathered. This work ran mainly for three or four midwinter months. I was a part of it for more than twenty years, and it led me to have an intimate acquaintance with the New York countryside, with many excursions into other states.

Something more than twenty-five years ago, Judge Berne A. Pyrke, then New York State Commissioner of Agriculture, suggested that I set down some of my observations, with especial reference to early agricultural machinery and the pioneer farm and household handicrafts. Some material that I brought together was published as Bulletin 203 of the New York State Department of Agriculture. This initial effort was poorly conceived in that it dealt mainly with what might be called the mechanics of the period, with scant attention to the social and cultural developments. The bulletin, which served as a beginning, has long been out of print, and it has seemed proper to retell the tale in expanded form and with much new material.

This book does not profess to be a general history of the pioneer period of New York State. It is rather an effort to collect and put into permanent form some record of the lore and the methods by which our forebears lived upon the land. It seeks to be a partial story of that now somewhat fabulous "homespun age" which began with our earliest settlements, which reached its fullest fruition during the generation following the Revolution, and which drew definitely toward its close about the time of the Civil War. That great national catastrophe retarded for a little the passing of the household arts, but its ultimate effect was to hurry and make more complete their extinction.

One disclaimer I wish to enter in very specific fashion. It is that neither in content nor in form of presentation has there been any attempt to copy the literary technique of the learned dissertation. The book is frankly of the earth, earthy. In it there is far more of memories, traditions, legends, and pioneer folk tales than of severely documented evidence. The very best that may be hoped for it is this: that in these soft days, it may lead to some degree of appreciation of the fortitude and high courage that animated the men and women who set up an enduring civilization in the tangled woodlands of northeastern America.

I have tried to preserve here—before the memory and traditions wholly perish—the farm life, the household handicrafts, and the rural occupations of that bygone era. Seventy-five years ago the available data were abundant and exact; seventy-five years from now even the voice of oral tradition will be dumb.

I have borrowed the title from one Horace Bushnell, Connecticut preacher, educator, and publicist, whose life span covered almost precisely the first three-quarters of the nineteenth century. Once, about the year 1850, speaking on the occasion of a New England town centennial and remembering his own youth and seeing it perhaps through a purple mist of years, he lingeringly and lovingly called it "The Golden Age of Homespun." I am sure that there was never happier phrasing for an epoch that had in it many of the things that make for romance, that nourished a civilization which bred men and women with many splendid qualities of heart and mind, and that is nevertheless an age that has passed on doubtless forever.

Living as we do in an era when specialized industrialism is supreme and when each separate requirement of life is supplied from some great manufacturing establishment, it is hard for us to realize that almost within the memory of living men—surely within the memory of a generation only once removed from our own—each farm community constituted an almost self-contained, self-supported, industrial, and economic unit. Time was when in every farmhouse was heard the whir of the spindle and the thack of the loom and when every farm family was fed, clothed, shod, sheltered, and warmed almost wholly from the products gathered from within its own fence line.

I confess to being one of those who view our agricultural past with an interest akin to affection. Prior to World War II we had come to believe that the twentieth century was the heir of all progress and of every civic and political blessedness. Today, when we are forced to invest so much of our money and our energy in the things of war, we are less certain that our progress is necessarily steady or inevitable. I submit that it is a proper subject for philosophical inquiry as to whether or not all of our boasted material progress has increased the total of human happiness. The last one hundred years have brought to the farmer as well as to the rest of the world an almost infinite progress in many ways. They have brought to us leisure and luxury beyond the dreams of our forest-conquering ancestors. They have even given us greater length of days, for vital statistics indicate that on the average we live longer than our fathers did. Men have learned to fly like birds across the sea, to whisper across a thousand miles of empty space, and to do things more marvelous than any fairy tales that were ever told.

In some respects we seem to have become like gods rather than men; nevertheless, I am not sure that it has brought us greater happiness, or content, or belief in ourselves, or hope for the future. Nor, to instance a very concrete example, have I any assurance that I find life fuller or more satisfying than did my own great-grandfather, who more than a century and a half ago walked these fields that I till and from them made a home for a patriarchal family. I believe the years of which I write were good years. I see no reason why I should offer sympathy to the man who in the first half of the century behind us owned in fee simple a good New York State farm.

I wish to acknowledge my indebtedness to a few special

sources of information. The New York State census reports for the years 1845, 1855, and 1865 afford data by which one may estimate the character and amount of statistical changes that have taken place in our agricultural crops and production. The Federal census of 1860 contains not only much of statistics but also condensed history of earlier periods. The *Transactions of the New York State Agricultural Society* are the year-by-year annals of the practices of the outstanding farmers during several decades. The early volumes of the *American Agriculturist* and the *Cultivator* offer firsthand knowledge of just what our farm people were doing from about 1840 to the end of the homespun era.

From my own father, whose life covered a long span and who had always collected and treasured traditions, I have received much. My father beyond most men cherished the past, and his memory was stored with twice-told, often inconsequential tales of his youth and the traditions he had received from others. Also he was one who trusted in the soil and fully and freely believed in farming as the best possible way of life. Then, besides my father, there were two old men who had something of the status of feudal retainers and who in some respects gave direction to my boyhood thinking. These ancient worthies were James Barker and John Brown. Both of them were distinctly older than my father, and Jim dealt with occurrences which, according to my youthful understanding, were distinctly prehistoric. Each man was a mine of half-forgotten memories and was always ready to rehearse old tales for my amazement and delight. I now suspect that, with their imagination fired by the eager attention of their audience of one, they were sometimes prone to draw the long bow. John died in 1904 at the reputed age of ninety, and doubtless I remember him better than anyone else.

Finally, I owe sincere and hearty thanks to a large number of correspondents to whom I have appealed through the columns of the *American Agriculturist*, and who have responded in many cases with long and detailed letters regarding early agricultural practices. Most of these letters have been from elderly men and women concerning facts they knew—sometimes by direct experience, more often by word of mouth.

Also, I should be remiss if I failed to express my indebtedness to Director Louis C. Jones of the New York State Historical Association and to his accomplished editorial associate, Miss Mary E. Cunningham, who have given me so much not only of counsel and guidance, but, even better, sympathy and encouragement.

JARED VAN WAGENEN, JR.

Contents

Illustrations

All starred illustrations are based on actual objects in the Farmers' Museum at Cooperstown.

The Golden Age of Homespun

CHAPTER I

The Homespun Age

TO ASK what, when, and where was the "homespun age" and what was the fashion of its appearance is surely a proper inquiry. It goes without saying, however, that it is impossible to assign to the homespun age a definite span of years and to fix a precise date for its close. Its beginnings were, of course, contemporaneous with the earliest colonization, and to some degree the age was universal in pioneer America. Its finest flowering was not along the coast but rather in the remote interior. The eastward-looking seaport towns, as soon as they became communities of any size, had always more or less regular contact with Europe, so that textiles, household goods, and certain articles which were the luxuries of those times came across the sea. After the first hard years at least, Boston, New York, Philadelphia, and some lesser towns never found it necessary to work out a self-sufficient economy to any such degree as was needful in the hinterland.

But with the westward-marching pioneer, it was a different story. Once he had turned his back on the scattered fringe of coastal civilization, every added mile of his progress decreased

1

the possibility of reliance upon the old homeland and made more imperative the cultivation of an independent self-sufficiency. By the time his advance had carried him a hundred miles inland on the thin picket line of civilization, which was continually pressing forward through the wilderness, he had almost completely severed his connection with the old home life and was confronted with the problem of creating the essentials of existence out of his own immediate environment. The extent to which the American pioneers, on the farm outside and within the home, developed a culture which was almost independent of the outside world is striking testimony to their resourcefulness and skill. The homespun age itself has gone down the stream of time, but there still remain enough tradition, records, and physical mementos that we may understand and—in our thinking at least—reconstruct the life of that period. While it is true that for most of us that era is finished and has become a tradition rather than a memory, it still in some degree survives in the backwaters of the Appalachian highlands.

In the corn belt country, and beyond, the homespun culture was never highly developed. There, colonization fell in what was a different economic era, and the railroads followed so closely after settlement that the homesteader had always ready access to the products of the new and dominant machine age. In the high adventure of conquering the West, there were glamour and romance and sometimes privations and danger, but there was never any development of that self-sufficient economy which prevailed for two hundred years in the back country beyond the Atlantic seaboard.

If it is at all possible to set any date or period and say, "Here the homespun age ended," then perhaps the Civil War is the most rational bench mark. In New York State it seems certain that by 1840 the old order was definitely going into eclipse, and yet in other ways the age held on bravely up until the cataclysm of that conflict. In my own family traditions there may be an answer to the question regarding the passing of that fabulous era.

My father was born on a Schoharie County farm in 1835, and surely he was born in the homespun age even if the twilight of that long day was coming on. When he was born, the little spinning wheel for flax and the big spinning wheel for wool were still a feature of every well-regulated farm home. He was almost a grown man before he wore any boots or shoes other than those made by the local cobbler from cowhides or calfskins grown on the farm and converted into leather at the community tannery a short mile up the road. In his boyhood, the lucifer match, if known in the country, was rare and costly enough so that if, by oversight or accident, the precious fire on the hearth was lost, one of two possible procedures was followed. Somebody might go to a neighbor's with an iron kettle in which to carry home "borrowed fire," or else very skillfully and patiently, he might strike steel on flint until a tiny spark would fall on prepared tinder which finally could be blown into a flame. These early memories of my father's boyhood are now a little more than a century old. Much of farm life was still carried on according to the old order; nonetheless that order was rapidly, almost suddenly, drawing to a close.

Thirty-six years after he was born, I followed him on the scene, and unquestionably I was born in the machine age. I never wore any clothing of domestic manufacture, save stockings and blue-and-white-striped mittens knitted from purchased yarn. The big spinning wheel and the little spinning wheel had gone to the attic, where they still remain. But I am glad to remember that traces of that bygone period still lingered. Dresses for women and shirts for the men were still made in the home, albeit of "store cloth." As long as my mother was with us, she made tallow candles and soft soap from farm fats boiled with lye from wood ashes. The household arts in general, and, most of all perhaps, the kitchen arts, still prevailed in my young manhood, and some of them survived until quite recent years. Memory still conjures up the pic-

Breakfast, 1887

ture of early winter morning breakfast eaten under the yellow glow of the hanging kerosene lamp, the meal built around buckwheat griddlecakes and sausage or some other preparation of meat. The farmer of two generations ago still had the traditional and abundant home-produced food, and my family carried on according to custom.

I am glad to boast that we have never wholly adopted the newer ways of life. When fall comes we still butcher a beef and two or three hogs, but I must confess that we are really giving ground, because we no longer practice all the traditional housewifely arts. We still make some liverwurst in the home, and there is a stone jar of our homemade salt pork down cellar, but we send the sausage meat to the local butcher to be ground, and we turn the hams, shoulders, and bacon over to him to be cured and smoked. When finished, we bring them home and put them in the deep freeze. This is a radical departure, but it seems all for the best.

The making of certain delicacies of earlier days—souse, rolliche (rolletje), headcheese, and pickled tripe—has become a lost art. So, too, the home preparation of corned beef and smoked, dried beef is on the way out. The facts would seem to add up to this: we still practice a certain ritualistic symbolism commemorating the homespun age when, down in our hearts, we know that era has finally departed. Somewhere during that span of thirty-six years which lay between my father's birth and mine, and which included the cataclysm of the War between the States, the age of the pioneer faded from the scene and the "new day" with new manners and customs took its place.

Any logical interpretation of the epic story of the American pioneer must not overlook the roster of that conquering army which slowly and yet irresistibly marched westward to take possession of the "undiscovered country." First, ever side by side, go a man and a woman. It was the man who advanced into the dark woodland to his selected spot and there laid about him with his ax so as to let the sunshine in on the forest floor and make an opening for his cabin. From this hearthstone as a base, he began his assault upon the wilderness that surrounded him. His armament was symbolized pre-eminently by the ax, the ox team, the plow, the scythe, the grain cradle, and the hoe. Always, along with him and surely not less heroic and honorable, must be reckoned his companion-in-arms, his wife. Her escutcheon should be graven with a house of logs and children playing by the door. Within there would be the great stone fireplace with its crane, the kneading trough for bread, the greater and the lesser spinning wheel and the barn-frame loom. Under this standard, and by these signs, she conquered.

And then, a little in the rear of this picket line of civilization, yet never far behind, went forward the second-line army made up of that wholly indispensable company who carried on the pioneer handicrafts. Any well-ordered pioneer economy demanded that the resources and skills of this supporting corps should be available to the settler within a distance not greater than could be traversed by a slow team, going and returning within the hours of a single day. These arts and handicrafts were so many and so diverse that it is difficult even to enumerate them without grave omission and oversight. But these at least are remembered: blacksmith, carpenter, cooper, cobbler, cabinetmaker, sawyer, miller, fuller, dyer, tailoress, tanner, mason, brickmaker, shingle shaver, wagon and sleigh builder, millwright, harness maker, charcoal burner, lime burner, and distiller. In some cases at least, these fundamental crafts were finely specialized. Doubtless it was true that in practice the building of baby cradles was properly within the field of the cabinetmaker, but making grain cradles (those indispensable implements with which most of the grain of the Northeast was cut until after the Civil War) was recognized as the peculiar province of certain highly skilled craftsmen. So, too, the boring of pump logs was another very special development of craftsmanship in wood.

How many persons were actually first-line farmers engaged in clearing the forests and making the beginnings of a home as compared to the number who were carrying on the rural handicrafts and in other ways ministering to the needs of the farm population? Unfortunately, the available census reports scarcely afford an answer to this question. The New York State census for 1845 was a great advance over earlier enumerations. It gives rather detailed information relative to vital statistics and also much information relative to crop acreages and yields, but it does not indicate the number of farm families nor does it enable us to know anything regarding the occupational activities of the communities.

The census of 1855 offers a large amount of additional data

which will help us to estimate what proportion of our people were farmers at that particular period. However, it must be remembered that 1855 falls not in the vigorous prime of the homespun age, but rather just at the time when it was going into eclipse. What we would like to know is what our social and economic life was like in the early years of that century when that age was still in flower.

These, then, are the two horns of the dilemma: if we go back into what was the real pioneer era, there are no dependable data available; if, on the other hand, we wait until there are at hand precise census figures, we find that the age is already drawing to a close. It would seem that the best we can do is to try to reconcile somewhat shadowy traditions with more recent factual findings. It has seemed to me that the most feasible approach to this question is to attempt a study of the social and economic make-up of some selected New York county in the year 1855. For this purpose, I have chosen Cattaraugus County. I make this choice because it was the last county of the state to be settled. Not until 1807 did a white man sound the ax within its boundaries. In 1855 its first cabin was not yet fifty years old. Each year new farms were being established, and during the decade preceding 1855 the inhabitants had increased by approximately ten thousand—evidence that the county was still in its vigorous youth. Some 1,252 families were reported as living in log houses—testimony to the extent to which pioneer conditions still lingered. What then was the occupational status of the people of this relatively new county in the earliest year concerning which we have any comprehensive knowledge?

To begin with, there were reported 5,441 farms and 6,855 farmers. This discrepancy is satisfactorily explained on the assumption that in many cases grown-up sons still at home would truthfully declare themselves to be farmers. There remained, however, a very large number—2,343 persons—who claimed some occupation other than farming. So, in 1855, in Cattaraugus

County, more than seventy-four per cent of all persons reporting their occupation classified themselves as farmers. This is not greatly different from an oft-repeated statement to the effect that in 1800 four-fifths of all our people were living on farms. If we disregard the seacoast cities and a few of the larger inland towns, this estimate seems entirely reasonable.

What were the occupations of these 2,343 persons whose vocations were reported as something other than farming? The most numerous were the 643 laborers—an all-inclusive group who probably in large part represented those who toiled for a wage in a day when the great mass of folk was self-employed. Next in numerical importance were the 267 carpenters—a disproportionately large group made necessary in a lusty young community whose whole economy, especially in construction, was rapidly expanding. There were 211 lumbermen, so obviously there remained a great deal of woodland in this recently occupied region of the state. There were 131 sawyers, indicating that a multitude of little sawmills powered by wooden water wheels were turning these woodlands into lumber. There was a husky crew of 160 blacksmiths. These modern disciples of Tubal Cain were possibly the most absolutely indispensable craftsmen of the pioneer era. The smith with his hammer and anvil and bellows came marching in almost abreast of the first settlers, and he was the last craftsman to disappear before the resistless impact of the machine age. Indeed, even now, here and there a blacksmith still stubbornly lingers on the scene.

There were forty-four cabinetmakers along with fifty-nine joiners, who wrought in order that the new county might be provided with tables, chairs, bedsteads, hope chests, baby cradles, and such other gear as the pioneer needed. There were 112 tailors and tailoresses (more commonly the latter) who often pilgrimaged from home to home among their clients in order that the farmer and his wife and children might be clad, although not so commonly in homespun as had been the case a half-century before. There were fifty-five tanners, probably representing almost as many tiny establishments in which local hides and skins were turned into honest leather for the use of the community cobbler and harnessmaker.

As representing other crafts, there were fifty-two coopers, thirty-nine millwrights, thirty-two shingle shavers, and eleven men who proclaimed their occupation to be gunsmith—this in a day when flintlock, muzzle-loading rifles, and shotguns were the almost universal firearms of the frontier. There were thirty-nine hotel and tavern keepers—a smaller number than might have been expected in a day when the main, long-distance roads were full of families with their possessions seeking new fortunes and new frontiers.

Relatively self-sufficient as the new country in that day may have been, the general country store was none the less a sound part of community economy. Always, there were certain items quite beyond the productive skills of the pioneer or his neighbors, and yet it seemed hardly possible to do without them. The storekeeper's list of commodities was very brief as compared with the almost infinite variety of today, but there was at least this incomplete assortment: brown sugar, molasses, salt, spices, tea and coffee, saltpeter (for curing meat), various kinds of salt fish either pickled in brine or dried, indigo, saffron and other dye-

Village Crossroads

stuffs, gunpowder, flowers of sulphur, and various drugs supposed to have medical value. Flour and meal were ground in the local mill for sale to those people unfortunate enough to have no land of their own. Many stores sold rum and other liquors. The voice of the temperance reformer had not yet been heard in the land, and most adult males drank on occasion.

From the very beginning, there had been some imported textiles, and by 1855 these were rapidly displacing the once very common fabrics of domestic manufacture. Then, too, the country store carried a certain line of hardware. Universally, there would be axes, grass scythes, cradle scythes, and tools for the carpenters, and for the blacksmith's customers there was always an assortment of nail rods, along with iron and steel rods and bars of various sizes. Whatever else might be lacking on the frontier, the blacksmith must not be without his stock of altogether indispensable raw material. No matter how new and primitive the civilization of the frontier may have been, life could not go on without certain utensils of kitchen use. In many places the immemorial craft of the potter flourished, and from his wheel came jars, jugs, and tableware.

In a few localities, glassworks had been established before the Revolution. In Massachusetts, New Jersey, and Virginia, iron ore had been smelted almost from the beginning of colonization. In New York State, Columbia County had an iron furnace as early as 1740, and before the close of that century iron was being produced at a considerable number of places in the Hudson River region. From the charcoal-fired foundries came teakettles, iron pots, and caldrons of various sizes, along with frying pans, griddles, and cast-iron bread pans. All of these things were a part of the staple stock-in-trade of the primitive emporiums which ministered to the necessities of the pioneer. It was a day when barter rather than cash played a large part in trade. Tallow, beeswax, dressed flax, and peltry were recognized as commodities that could be transported long distances and that had a dependable cash value if only they could reach the far-off seaport towns which traded with the world beyond the sea. So it was that in the County of Cattaraugus in the year 1855 there were 137 persons who described themselves as merchants, which in most instances indicates proprietorship of one of the little general stores to be found in the crossroads hamlets.

There remains the consideration of the professional life of the countryside. It is a striking commentary upon the primitive conditions prevailing in 1855 that in all Cattaraugus County, geographically an extensive area with almost 40,000 people, there were only six men who professed to be dentists. Dental work as we understand it was practically nonexistent; aching teeth were pulled by the physician. As a matter of record, up until a time within the memory of many living men, most of the older country doctors pulled teeth as a regular part of their daily activities.

However, even at this date, the three traditional learned professions were pretty well represented. Always in our state, after the most advanced wave of settlement, the wooden courthouse was soon erected wherein was enacted the ancient ritual of English jurisprudence. Always in these early days there was much activity in the conveying of land titles, and not infrequently there were actions-at-law to determine disputed ownership, so that there was ample opportunity for the accredited disciples of Blackstone, to say nothing of the self-taught justices of the peace who in their field did not hesitate to render judgment. Considering all this, it is remarkable that in the entire county there were found only thirty-three men who declared themselves to be lawyers.

The frontier from the beginning had need for physicians. Daily life was hard and dangerous, and accidents were frequent. There was little knowledge of sanitation or hygiene, and both typhoid and tuberculosis took their toll of the rural neighborhood. Also, with a frequency which today seems incredible, new babies were being added to the community. So it was that the followers of Hippocrates flourished, and in Cattaraugus County there were seventy-four men who confessed—or averred—that they were physicians. Be it noted that there were more than two doctors of medicine for each counselor at law. Apparently popular judgment deemed it preferable to live by men's ills rather than by their quarrels.

It has always been characteristic of life in America that the organized church came soon after the first wave of advancing settlement. Even in 1845, when Cattaraugus County looked back less than forty years to its first white resident, there were already thirty organized churches, all Protestant. In 1855, ten years later, when fairly detailed statistics are available, the list had grown to sixty-one, of which four were Roman Catholic. At the same time there were in the county sixty-four men who gave their occupation as clergyman.

Occupational statistics for Cattaraugus County, or for the state as a whole, are pretty consistent until one considers the matter of schools and schoolteachers. Here there appear certain discrepancies in the figures such as can be explained only by surmise and assumption. In 1855 there were in the State of New York more than 11,000 rural school districts, to say nothing of the cities. But that same year there were in the whole state less than 10,000 men and women who told the census man that their occupation was teaching. Considering that almost 1,300 of these were in New York County—not to mention substantial numbers in upstate cities—one cannot but wonder in what fashion the rural schools were provided with teachers. Cattaraugus County, having served as "exhibit A" in so many cases, may again be questioned concerning this matter. The answer is startling. The census for 1855 reports a total of 242 schoolhouses, but only ninety-nine persons confessed that their occupation was teacher. Approximately the same conditions were found in many other counties, especially those which in date of settlement and development were nearest to the pioneer age. Two explanations may be offered. Under the Common School Act of 1812, the state had been divided up into more than 11,000 school districts. It is probable, however, that some districts, although officially set up, were not yet functioning. Also, this was before the day of departmental certification of qualifications, and any person desiring to teach school was at entire liberty to do so, provided only he or she could find some trustee willing to employ him. So it was that almost anybody who so desired could teach. We know that, occasionally at least, a man exemplified the twin dignities of clergyman and pedagogue. Sometimes wise and understanding housewives and mothers of families became teachers because in their girlhood they had enjoyed a term or two of instruction in some primitive "academy" back in New England where, as everybody knew, scholars were made. In many cases, the schoolmaster was a farmer who somehow or other had picked up a working knowledge of "reading, writing, and the rule of three" and thereby was incontestably qualified for his profession. These farmer-teachers might consider themselves farmers by vocation but were happy to assume the avocation of teaching in the winter when the schools were functioning and the farm work was not too pressing. It goes without saying that very, very few of these bygone instructors would have been able to exhibit any engraved parchment setting forth their academic attainments. Nevertheless, some of them doubtless had at least some small spark of that divine fire which dwelt in Mark Hopkins so that, as the tale runs, "he on one end of a log and a student on the other . . . there in the wilderness was set up a great university." Also, it is to be feared that some of them justified the declaration of the Cornell blacksmith (and G. B. Shaw), "Them as can does and them as can't, teaches."

But, whatever may have been the capabilities of teacher and pupil, the architecture and equipment of the rural schools of the homespun age were crude and primitive beyond our modern conception. As late as 1855 we had in New York State 313 log schoolhouses, and three of these were approved, or at least tolerated, by the Department of Education until after the turn of the present century. The rural school year of a century ago was very brief—often not more than twenty weeks. These are some of the reasons why schoolteachers as a class had far less of professional consciousness than was held by lawyers, physicians, and clergymen, and it explains why the number of avowed teachers was very much less than the number of operating schools.

CHAPTER II

The Setting of the Stage

IN THE foregoing chapter an effort was made to show something regarding the pattern of life in the rural communities of New York State at a period still within the homespun age, albeit that era was very rapidly drawing to an end. It was based upon a sketchy survey of the agricultural, industrial, and professional setup of a single county, selected primarily because it seems to have been at this date most representative of the pioneer epoch. Conditions there were still raw and primitive, while the Hudson Valley had already attained a comparatively mature civilization. Because the county of Cattaraugus happens to be located in that hinterland of our state which was latest in settlement and development, it may be assumed that conditions there in the year 1855 indicate something of the setting of the stage on which was enacted the long, patient—sometimes glowing—pageant of the homespun age. The early beginnings of the story are shadowy and fragmentary. Even as the age neared its end there are parts of the picture that lack authentic detail. For example: not until 1814 did the State of New York enumerate the farm animals, and it was 1845 when first were counted the crop acres and the total agricultural production. Before these dates are the dark ages.

I believe we do not generally realize how far the state had developed agriculturally by that milestone year 1845. The harsh rigors, dangers, and privations of the earliest pioneer years were gone forever, and everywhere were to be found the beginnings of a secure and substantial rural civilization. Most of the strictly agricultural regions of the state had a population even larger then than today, and eager settlers were filtering into the narrowest valleys and occupying even the windswept hilltops.

The state had long been divided into school districts, their boundaries in most cases identical with those of recent years and their size determined by what was regarded as the possible daily radius of walking of a five-year-old child. In the primitive schoolhouses a now-vanished race of schoolmasters who had never heard of child psychology or the principles of pedagogy taught at least the three R's more soundly and thoroughly, some oldsters will insist, than now. In the courthouse of every county, the stately ritual of the law went its orderly way, and justice was done, certainly more expeditiously and perhaps more surely than in our time. In every considerable hamlet was seen the church spire, and crusading scouts of Zion were founding new churches with a zeal which may have been apostolic but was often misdirected. Denominational controversy frequently reached a bitterness that is almost unknown today, and many a church was planted as an expression of denominational triumph rather than for the glory of God.

As has been said before, the census of 1845 constitutes our earliest findings concerning what we have come to call vital statistics. Rather surprisingly, perhaps, this report offers fairly comprehensive data concerning the birth and death rates, together with the relative numbers of various age groups, as well as a good deal of information pertaining to the marital status of

The Schoolmaster

our population. A very casual inspection of these records affords convincing evidence that the stork hovered over the home with a persistency almost unknown in these degenerate days. The fecundity of the American rural community in the first half of the nineteenth century was a phenomenon that amazes the student of present-day population trends.

The simple standards of living, the feeling of security arising from the abundance of home-produced food, and the thronging tasks of both home and fields made children an economic asset rather than a liability, and they were welcomed as fortunate contributions to the family welfare. Then there was the added possibility that these hearty folk, very sure of themselves and their future, unconsciously felt it a duty laid upon them to people an almost untouched continent.

For confirmation of these general statements, we may once again turn to the records of Cattaraugus County. In 1845 there were living in the county 3,991 wives between the ages of sixteen and forty-five, and the previous year there had been born to them 1,078 babies. More important however, is the overplus of births above deaths. During that same period there had been 341 deaths in the county, so that every time the community filled a coffin, it filled more than three cradles.

Further testimony concerning the birth rate of the American pioneer may be gathered from *The Beginnings of New England*, by historian John Fiske of Harvard. He observes that a population of about 26,000 had been planted in New England by the year 1640, after which, owing to changing political conditions in Old England, there was only limited and occasional emigration for more than a century. Surely, those original 26,000 were relatively only a little group of Englishmen in a vacant land, but "they continued to multiply on their own soil for a century-and-a-half in remarkable seclusion from other communities." The facts are that by the period of the Revolution, New England was agriculturally overcrowded, and hardly had peace been declared before the Yankee was on the march seeking new lands and new fortunes beyond the far-off Fort Stanwix treaty line. There was what seemed an overflowing, possibly inexhaustible, reservoir of people which spilled across the Hudson, settled much of New York State, and ran on into Ohio and Indiana and the states beyond until the second or third generation of the Argonauts founded Portland and Salem in the far West and named them in memory of the old towns "back home." It has been stated that time was when, by natural increase alone, without benefit of immigration, New England's population doubled in number each period of thirteen years. What may have been specifically true of those particular states applies to pioneer America and the homespun age in general. If a comparable birth rate had been maintained until now, we would have discovered that the famous doctrine of the

Reverend Thomas Malthus was something more than an academic hypothesis, and ere now, we here in America would have been compelled to display the "Standing Room Only" sign.

Because of the teeming families the rude schoolhouses were full of boys and girls, and on Sunday the churches were crowded with people—not necessarily because men were innately more religious than now, but because the church was the social center, the common meeting place and the clearing house for neighborhood news and gossip. In addition to this there was, especially in communities of New England extraction and hence of Puritan traditions, the compelling force of public opinion. The man who did not go to church felt himself outside the pale of community approval. Still it will hardly do to picture too idyllic a religious life even in those somewhat fabulous times. There are still tradi-

The Horseblock Class

tions which say that while the church may have been full of worshipers, there was also the "horseblock class" which convened outside or under the church sheds and told dubious stories or discussed horse trades until the more devout were released by the preacher. If today we feel that we must bemoan a less rigid standard of church attendance, we may at least remember that there is no longer the force of public opinion and that today men go to church or stay away as they see fit.

I think it certain that by 1845 the household and rural handicrafts had already entered upon their long decline, but concerning these facts we can have no exact data. Life was still exceedingly primitive, and the farm home was still self-centered and self-contained, but industrial villages were springing up, railroads were being projected and constructed with enthusiastic energy, and a new era was about to be ushered in.

The decade 1840–1850 was a fortunate period for the New York State farmer. The New England and North Atlantic states were already well occupied agriculturally, and any rapid development of their resources was no longer possible. Indiana and Ohio were shipping wheat and wool, and driving fat cattle and sheep to the seaboard; but not until 1847 did the frontier town of Chicago see its first railroad train, and the Mississippi states had as yet hardly embarked upon that spectacular career of expansion which was destined forty years later to swamp the agricultural markets of the world. I believe that about this time we had a definite congestion of population. Farm labor was in almost unlimited supply, and wages were low—pitifully low, perhaps. Prices of agricultural products were relatively high. The man who owned a hundred acres of land in fee simple felt that his was a goodly heritage.

The impulse toward scientific agriculture was spreading throughout the land. Our farmers subscribed for and read the *American Agriculturist*, the *Cultivator*, and Moore's *Rural New Yorker*, which were examples of an enterprising and worthy agricultural journalism. Men were beginning to discuss and agitate for a New York State College of Agriculture. The New York State Agricultural Society was holding annual meetings and publishing its *Transactions* in which the best farmers of the state set down their experiences. John Johnson, the Scotch emigrant, had introduced the practice of tile drainage on his farm near Geneva, and his example was being very widely copied in large portions of the state. In 1848 Johnson's friend and neighbor, John Delafield, imported the first tilemolding machine from Scotland. Here and there, successful farmers were bringing in breeding animals from the best herds and flocks of Europe, and these and their progeny were being exhibited at the New York State Fair. Cyrus McCormick of Virginia was perfecting his reaper, and Burral and others were rapidly improving the threshing machine. Everywhere there was eager interest in better agricultural practices. In our agricultural literature there was nowhere a note of despondency. The farmer felt very sure of himself and his future.

In the year 1845 we had a million and a half hogs—considerably fewer than in 1840, but a larger number than was ever again attained. We had nearly six and one-half million sheep. What a pitiful remnant of this great host now remain! We had almost exactly the same number of cattle as now, but it is only fair to say that the cow of today is a better animal and that she is far more liberally and skillfully fed so that our daily production has greatly increased.

Population comparisons between then and now are not easy to make. The township was and still is the unit of enumeration, but it is impossible to reckon the changes due to the decline of the hamlet and the rise of the villages that have grown up at the railroad stations. Unquestionably, there were more people living on the farms in 1845 than now. In most cases rural population reached its maximum in the census of 1860. The Civil War greatly accelerated many social and economic changes, which in any case were bound to come. By 1870 the innumerable hamlet industries were definitely in eclipse, never to return. The big town and factory system with mass production had come, and the never-again checked migration from the country had begun. Apparently the New York State farm held its own until 1880 when we had more farms than ever before or since, although the maximum of acres, including those in hay, did not come until twenty years later. In any case, the number of farmers and the number of tilled acres have been steadily declining since the first of this century. How long it will continue, no one knows.

It is good to record that there are indications looking toward the hopeful conclusion that the decline in rural populations has at length reached its nadir and that a trend almost unbroken for three-quarters of a century is at least arrested and perhaps re-versed. If this be indeed true, it bodes' well for the social and economic welfare of the countryside in the years ahead.

CHAPTER III

The Pioneer Goes Forward

IN OUR historical thinking and our patriotic speechmaking we have built up an almost standardized conception of the winning of the wilderness. It may be stated as "the irresistible, all-conquering march of the American pioneer." That is a fine, impressive, mouth-filling phrase. For myself, I like to use it, or one very much like it, now and then. Somehow it suggests an exultant army marching briskly and confidently toward the Pacific Ocean with drum beating and banners waving, brooking no delay.

Perhaps, at this distance in time, we fail to realize the difficulties and the vastness of the task involved in taking possession of the forest reaches behind the Atlantic seaboard. Let us inquire concerning the speed and the ultimate span of this monumental march. New York State was unique and fortunate in that the Hudson River offered a tranquil, easily navigated water corridor stretching a hundred and fifty miles into the interior. That is why the colony at Albany is as old as the first settlement at the mouth of the river. Yet, in spite of the great advantage afforded by such a good jumping-off point, the settlement of the hinterland went forward with what must be regarded as painful slowness. In 1714 when first the Dutch and later the English had been established at Albany for a full century, Schenectady, the only sizable town not on the river, was a little more than fifty years old. The very recent German settlements in the Schoharie Valley, less than forty miles from the Hudson, represented the extreme western outposts of the white man's culture. Everywhere beyond, the writs of the Iroquois still ran, and these weather-beaten tribesmen met a little contemptuously, perhaps, the wandering fur traders who from time to time penetrated their domains to barter for peltry. That could hardly be regarded as breathtaking progress over a hundred years. If we reduce it to miles, it appears that during this period there had been, on the average, an annual advance of considerably less than one-half mile.

However, during the second century of occupation, the pioneer did a good deal better. By 1814, when Albany was two hundred years old, there was a little village of log houses where Buffalo Creek emptied into the lake, and there was no one of the present counties of the state but had its settlers. During this period, civilization had crept forward through the forest at the average rate of three miles each year. Settlement was a constantly accelerated movement gathering momentum with each passing decade.

It would be vain to try to draw a line on the map which would precisely define the advancing front of the army of settlement and to say that here in a certain year was the frontier of agricultural civilization. In any case such a line would be wavering and uncertain, with many bulges and salients. It would be safe however, to say that by the decade 1830-1840 the pioneer army had at length won through the eastern woodlands which had held it back so long and finally stood on the edge of the region we now call the corn belt. Before them stretched an utterly different country, a land that was level, stone-free, almost fabulously fertile, and blessed with a kindly climate. Best of all, as it must have seemed to them, it was covered not with dark forests, but protected only by the immemorial turf. There was a vast area of it, the very heartland of a continent constituting the richest agricultural prize ever to be exploited in all the long history of the race.

Then, too, by 1830 the steamboat was a frequent sight on the Great Lakes and on some of the western rivers. In 1847 Chicago became a railroad terminal. The machine age was about to come to the help of the pioneer.

So it was that when he started on the final stage of the conquest of America, the way was relatively easy. For more than two centuries he had been slowly and painfully pressing his way through the tangled wilderness with incredible labor and at a pace which seemed to take no account of time. Now with the treeless prairies before him and with steam come to his aid, he literally romped forward. The rapidity of the conquest of the remainder of the continent seems almost incredible. In the twenty years between 1840 and 1860 the population of Iowa increased eighteenfold, and it was predominantly an agricultural development. In the five years from 1855 to 1860, Kansas grew thirteenfold, and in the single decade 1880–1890 that lusty young state added more than a million inhabitants. In 1846, which was three years before the California gold rush, men and women in ox-drawn wagons threaded the passes of the western mountains and stood beside the Pacific Ocean. Surely, it was a long, long trail that led back to Jamestown and Plymouth and Albany.

In 1840, the land-hungry farmers of America had barely begun their possession of the seemingly boundless acres of the public domain. However, by 1899, when the Cherokee strip was opened up for homesteading, it was recognized that then almost the last of the really fertile and desirable lands had passed into private control. It marked the end of a long era—the era of governmental distribution of free land. There had never been anything like it before, and it would seem that anything like it can never be again. It might be a safety valve for American unrest and turmoil if we still had an agricultural frontier.

CHAPTER IV

The Clearing of the Land

CONCERNING the herculean (there seems to be no other adjective so fitting) task of clearing the land, we have only the sketchiest and most incidental contemporary accounts. It was by no means a wholly illiterate age. Men kept records: court records, vast archives of county clerk records, and town board, school district, and church records. They wrote letters and kept account books, and not a few kept diaries. The operations, however, of clearing the land were so much a part of daily life, so commonplace and self-evident, that they never once thought to set down the manner in which this task was accomplished. Since there are no written records, I would like to believe myself the heir of enough farm and family traditions, so that I may set down at least the outlines of the practical technique of subduing the forest here in east-central New York, a procedure which doubtless would apply to almost any part of our old Northeast.

To advance upon a piece of tall timber and prepare the ground for the first crop, the pioneer needed, besides his own high courage and skill, three—and only three—tools, to wit, his keen ax, his firebrand, and his trusty ox team. Man has always found the ax the most necessary of tools. The Stone Age man searched until he found a flint fragment somewhere near the proper size and shape, happily broken to a keen cutting edge, and, inserting it between the two halves of a cleft stick, he bound it in place with rawhide thongs and so had a better thing—an ax.

The Pioneer and His Tools

However, long before the first colonists came to America, men had iron axes with an inserted cutting edge of steel, pierced with an eye so that a handle could be inserted. Doubtless, if he were given iron and steel with which to work, a passable ax was quite within the skill of the old-time smith, who was accustomed to hammer out almost anything that his client demanded. There remain abundant mementos of the premachine age in carpenter's tools in wide variety, including the broadax and the adz, but early specimens of the common wood chopper's ax are almost nonexistent. The explanation for this is that the ax was not a long-lived implement and had in it very little of the heirloom character. Being in almost daily use, it needed frequent grinding and hence wore out quickly. In those days no bit of iron was lightly cast aside. I know that it was standard practice to weld together two worn horseshoes and so make one new shoe, and doubtless a resourceful smith could always devise some useful purpose for a worn-out ax.

We must remember that the things that survived the generations were the implements that never wore out but simply became obsolete and were kept because they were "too good to throw away." Those early axes, products of the smith's skill, surely lacked the finish and polish of the modern die-forged implement, but I know of no reason why the smith should not have produced a serviceable tool. Indeed, the prodigies of tree felling that the pioneer performed bear testimony to the fact that he must have swung an at least fairly efficient blade.

I do not know that it is anywhere written in the books, but tradition and the fading memories of old men in the generation preceding my own bear witness to the fact that the subduing of the forest of the Northeast was accomplished almost wholly by the ax. The saw is, of course, an old tool as used by carpenters and also in sawmills, but the common crosscut saw, which has for a good many years largely displaced the ax in all lumbering operations, was apparently unknown in the days when New York State farms were being chopped out of the wilderness. Early log cabins show the marks of the ax rather than the saw, and my father used to speak of the cross saw as a distinctly new tool when it came to the farm sometime in his boyhood memory. It is my belief that the ax was made in many places, doubtless sometimes by smiths who never thought of themselves as manufacturers of this implement. Burtonville, in Montgomery County, and a tiny, unnamed hamlet in Albany County, are traditionally spoken of as having once on a time had "ax factories." Probably this meant that two or three skilled workmen stood beside their bellows and charcoal forge and beat out the weapons which were distributed in their local bailiwick. It may well be that the universality of the ax and the laggard introduction of the saw came about because the former was within the craftsmanship of many local smiths whereas the saw was utterly outside the range of their skill. This again is surmise and unproved.

But, in any case, it must not be forgotten that in the pioneer period—meaning the days in which our New York State farms were being hacked out of the forest—it was the ax rather than the saw that laid low the unmitigated wilderness.

I thrill to that fine phrase, "sounding the ax." Its muffled, rhythmical beat in some never-before-touched forest glade constituted a sort of official announcement that at last the white man had come to possess the land. Anyone who by experience has had intimate contact with the toil involved in subduing the woodland cannot but wonder at the physical hardihood and the high hope, faith, and courage which could inspire a man to lift his puny arm in warfare against the primeval growth of trees which once blanketed the state. Seeking an explanation, I have accepted the one offered now and again in the Old Testament accounts of the conquest of Canaan by the children of Israel: "There were giants in the earth in those days," (Genesis 6:4).

By what may be called hereditary memory, I am not far removed from this era. My maternal grandfather was born in 1814 in a log house in the Town of Carlisle in Schoharie County. His father, John McNeill, had come thither a few years earlier and taken title to a hundred acres of land which was untouched wilderness. In this hard-bitten, Scotch Presbyterian pioneer was, after all, the soul of a poet, because many years later when he was old and about to die, he gave commandment that he be buried at the exact spot where he cut down the first tree to make an opening for the cabin. Also, I have received much from my father because in his youth he was familiar with the talk of aged men whose experience included active participation in clearing portions of this farm. To me, these self-legendary, word-of-mouth traditions have something of the romance described by Wordsworth in "The Solitary Reaper"—of

> old, unhappy, far-off things,
> And battles long ago.

This, then, was the fashion of the settler's advance upon the forest that surrounded him. To make a beginning, he went to some chosen spot, chopped down a tree, turned where he stood, and chopped down another and then another. He cut them down and nothing more, careless of how or where they fell but watchful to avoid lodging one tree in the branches of another still uncut, because if this happened it added greatly to the difficulties and more especially to the dangers of his task. It was hard, grueling, relentless work pursued with grim determination until some acres of woodland lay prostrate, a tumbled confusion of tangled and crisscross trunks and uplifted branches infinitely more difficult of penetration than before. There are parts of the country where such a tract of felled timber is called a slashing, but in our local speech it was "foller"—probably a variation of the old English term "fallow," although that word as defined has a considerably different implication.

The ideal time to cut down trees was in early summer when they were in leaf. These leaves did not fall off but clung to their twigs and evaporated the moisture from the branches, and they were a great advantage in securing a good burn in September. The practice was that this new-cut woodland should lie all summer as it fell, drying out and getting ready for the next phase. Then some bright, dry, breezy day in autumn the settler's second tool, the firebrand, came into use.

If conditions were favorable, and especially if there was a considerable proportion of pine and hemlock, the fire would sweep across the foller in a line of swirling smoke and crackling flame, consuming the leaves and underbrush and smaller branches and leaving behind the fire-blackened stumps and trunks along with many of the larger limbs. When the fire had died, the settler went forward once again with his trusty ax and, lopping off the limbs that had survived the first burning, threw them into piles to be burned at some later date, and cut the trunks into lengths that might be dragged (snaked) by a yoke of oxen. It is not a feasible operation to burn a single half-seasoned log, but if the logs are rolled up in a heap of a score or half a hundred, a fire once started will grow into a raging pyre that will consume everything. In the earliest days of settlement on the first thin picket line of civilization, everything was burned as less than valueless. A little later when primitive sawmills were established and some sort of market for lumber became available, some of the very choicest oak and pine might be sorted out and saved from the burning; but such salvage does not alter the fact that by far the larger part of the splendid woodland that once covered New York State was cut down and burned in roaring follers or hot and glowing log heaps.

We should not for a moment be censorious of our forebears because they were so far removed from what we call conservationists. During two hundred years of New York State development, it was trees, trees, always trees, that stood between the pioneer and smooth fields, and he came instinctively to feel that the man who in any possible fashion destroyed a tree was a public benefactor deserving well of his fellows. Not until rather recently have we come to appreciate the sentiments found in George Pope Morris's poem:

Woodman, spare that tree!
Touch not a single bough!

If the settler had any land already cleared for crops, the simple, best, and easiest way to handle the newly burned fallow was to fence it, use it for pasture, and let the years take care of the stumps. Fortunately, cattle are fond of browsing the young growth of deciduous underbrush so that land where cows run all the growing season has no chance to revert to forest. When this course was possible, eventually with the years the stumps moldered away and the land was more or less ready for the plow. The length of time required thus to get rid of stumps varied greatly according to the species that made up the original woodland. Birch, beech, sugar maple, and basswood stumps are short-lived and make very little trouble after a half-dozen years. Elm, ash, hemlock, and soft maple persist longer, perhaps a dozen years. White oak and chestnut will cumber the ground for at least a quarter of a century, while the life of a big first-growth white pine stump must be measured in generations rather than years. No pioneer who cleared the land could ever hope to live long enough to see them disappear of their own accord. So it was that some of the best farmers in later years, when some measure of prosperity came to them, did a great work in grubbing out these stumps and setting them up to form about the strangest type of fence ever devised. Until relatively recent years these stump fences were a familiar feature of the landscape in many regions of New York State.

Of course, the settler was happy when he was fortunate enough to get a "good burn," meaning thereby, a fire that made a clean sweep and left behind as little as possible to cumber the ground. However, there was a belief that a burn might be too good. It is possible that in a very dry time a hot fire might eat down and consume the leaf mold accumulated through the centuries. If this really occurred, it was, from the standpoint of soil fertility, a major disaster. In my father's time, this belief was abroad in the land,

and he held it by tradition rather than experience. At any rate, on an adjoining farm there is a field which somehow or other, even today, is not as good as the land that surrounds it. More than once my father discussed this circumstance and then he would add, "Uncle Willis told me that when that field was cleared off the foller got an overburn."

Stump Fencing

This burning of follers was in reality a sort of regulated conflagration which over a period of two centuries eventually embraced the greater part of New York State, and, for that matter, nearly all of the woodlands of eastern America. On a bright, dry day in September or October a man might lift his gaze to almost any quarter of the compass and there—sometimes close at hand, sometimes beyond the first horizon—note the billows of smoke which betokened the fact that some settler was pushing back the forest so that he might increase his agricultural acreage. It has been a long time—two full generations at least—since such a sight was at all commonplace. What we are witnessing today is landclearing in reverse. It is a common estimate of the Department of Conservation that one-sixth of the state, say three million acres, land that once had known the plow, is now quietly slipping back into the forest from which it was once wrested by such heroic labor.

CHAPTER V

The Glorious Ox Team

AT THIS distance it is hard for us to conceive just how the pioneers managed to survive the first year following their arrival on the extreme frontier. Often they endured harrowing privations. Judge William Cooper, father of the novelist and sponsor of the Cooperstown settlement, has left a graphic picture of the hardships of their earliest years. In the winter of 1788–1789, when the infant colony was about a year old, it came near to perishing of hunger. Cooper indicates that they had expected to receive food from the nearest more-developed settlements in the Mohawk Valley. The Valley crops, however, were short and the demand unusual, so that there was nothing available from this source. The Judge, who seems to have served as a sort of father to his distressed people, distributed among them several loads of provisions originally intended for his own family and workmen. In addition he obtained from the legislature an emergency appropriation to purchase 1,700 bushels of corn, which were packed into the community on the backs of horses. The settlers were eventually reduced to dining on the edible and perhaps nutri-

34

tious, but surely malodorous, bulbs of the leek or wild onion. With April, however, came a providential and miraculous draught of fishes. Some migratory fish (the Judge calls them herring) ascended the Susquehanna in incredible schools. The settlers, using rude handmade nets of woven twigs, waded into the stream and threw fish out on the banks by thousands so that "within ten days every family had an ample supply." He adds that they had on hand an abundance of salt. The food shortage was thus happily surmounted.

Pioneering in that eighteenth century was at best a desperate adventure. If the newcomer's holding, as must usually have been the case, consisted wholly of virgin wilderness, it would be at the very shortest a full year before his farm could make any important contribution to the family food supply. The best he could hope for was that a foller chopped down in spring or early summer might be burned over in September and winter wheat scratched into the still warm ashes. Then if all went well, the next July he might have a patch of wheat with the blackened stumps standing like tiny islands in a golden sea.

In some of the less fertile regions of the state, and particularly in those with inherited New England traditions, it was corn rather than wheat that held the larger place in pioneer economy. So, while part of the settler's very first clearing would probably be sown to wheat, it was certainly expected that part of it would be reserved for corn, and the next May it would be planted with the Indian grain, not in regular, orderly rows, but in haphazard fashion wherever room could be found among the stumps and roots. Such cultivation as it might afterwards receive was with the hoe. Appalling as must have been the difficulties of making a home in the new country, there were at least two favorable factors. For one thing, in the first few years following the clearing there was no immediate need of fertilization, because the long-accumulated leafmold and the minerals in the ashes took care of that. The other compensation was that fortunately the pioneer did not have to contend with any such problems of weed control as vex the modern husbandman. Virgin soil was practically free of weed seeds. Strangely enough, nearly all of the troublesome weeds of the north-eastern countryside are introductions from Europe.

It is evident that even with the greatest resourcefulness and industry, the settler for at least a year after his arrival must depend upon such food as he could bring with him, supplemented by supplies obtained from the older communities in his rear. Now and then, but surely only rarely, some happy chance attended his choice of a location. A case in point was the settlement of the Schoharie Valley in 1712. This, as I have noted, represented the westernmost advance of the white man at that period. This colonization was not the result of gradual infiltration, but rather it was a deliberate, planned mass migration well beyond the limits of settlement. A large and homogeneous company of German Palatines, probably about 600 in number, marched into the wilderness to take possession of a tract of twenty thousand acres already purchased from the Indians by an agent acting in their behalf. They were particularly fortunate as their holdings embraced some of the loveliest and most fertile river bottoms in the state and several areas of land already cleared by the Iroquois. Because of this, they were able to sow some wheat in the autumn following their arrival. The resultant crop was reported to have been almost unbelievably abundant. In the earliest years of the settlement they had no draught animals and these old Indian fields were dug up with broad, heavy hoes. In doing this they found many groundnuts, which materially eked out their scanty food supplies. Presumably, these groundnuts were the edible tubers borne by a climbing leguminous plant (*Apios*), common along rich river bottoms in the eastern United States. Unquestionably, the white man was frequently indebted to the Indian for knowledge concerning the unsuspected food resources of the wilderness.

It was quite in keeping with the times that within a week of the arrival of the Schoharie pioneer, four babies were added to the new community. One imagines that there was something akin to a shout of triumph in the wilderness. The fact that the full Christian and sire names of these four have been preserved lends an especial air of authenticity to the story.

The chronicler of these early years of the Schoharie settlements was one "Judge" John Brown. By vocation he was a wheelwright and farmer, and by avocation justice of the peace and associate judge of the Court of Common Pleas for the County of Schoharie. Also by common consent, he was "man of affairs" in general. Withal, he possessed the instincts of the antiquarian and historian. Born in 1745, he lived close enough to those times so that his memory was the repository of much word-of-mouth testimony, both direct and traditional.

The "three Ithaca pioneers" seem to have enjoyed a happy fortune similar to that of the Schoharie Valley settlers. Reaching the site of Ithaca in the autumn of 1787 they happened on "the ancient maizelands of the Cayugas." The story runs that they cast seed and then made the trek back to Kingston, returning the following spring, along with their families, in time to build log houses and harvest their wheat.

Farther west in the Genesee country certain fortunate settlers benefited by possessing the corn and bean fields of the Senecas, but such good fortune was of course the rare exception. It was more often the rule that only after backbreaking, perhaps heartbreaking, toil might the trailmaker hope to sit under his own vine and eat the fruit of his own land.

During the first generation following the Revolution, many a Yankee left settled countryside and with his young wife and their few belongings in an ox-drawn wagon set a course for "The Purchase," west of Geneva and beyond the Massachusetts Preemption Line. In those days such a long migration meant the severing—often permanent—of old home ties. It was a day when men and women dauntlessly fared forth on great adventures, but they did not travel casually as today. It seems certain that for most of them it was a permanent exile and that never again did they see the lovely reaches of the Connecticut Valley or the stone-strewn pasture fields of the land where they were bred.

So far as cutting down the trees and burning over the foller is concerned, the pioneer might, if necessary, proceed single-handed, but when it came to drawing in the blackened trunks and piling them in heaps for the burning, he was confronted with a task beyond his own unaided efforts. Then it was that he availed himself of the universally accepted custom of the new country; he summoned his neighbors to a "bee," in this case, specifically called a log rolling. Whatever may have been the shortcomings or faults of pioneer society, it at least laid much stress upon the fine virtues of community good will and mutual helpfulness. So the other settlers within reach came with their

Log Rolling Bee

ox teams and skidded, or snaked, the charred logs to convenient spots where the united efforts of strong and willing workers rolled them up into heaps ready for the burning. There is no doubt that our modern term, log rolling, is derived from this particular pioneer activity. Once it denoted an entirely praiseworthy, co-operative community effort.

Doubtless, some day there will arise a great creative artist who in worthy fashion, on mural or canvas, or perhaps in bronze, will depict the epic labors of the pioneer as he cleared the land. When this is done the settler's animal helpers and co-laborers will be drawn, not with flashing eye and arching neck, prancing and rearing like those steeds on which conquerors and kings have been mounted in all history, but they will be great patient brutes with hooves that grip the earth, mighty shoulders, heaving flanks, and drooping heads, placid eyes, and spreading, gleaming horns— the glorious ox team.

I believe there is small danger of exaggerating the place occupied by oxen in taking over the new country. For more than a generation following the close of the Revolution, the hordes from New England in unbelievable numbers continued to pour into the Promised Land of New York, and according to universal tradition, the favorite vehicle of the invasion was the ox-drawn wagon or in some cases the two-wheeled oxcart.

Unquestionably, we fail to realize the almost indispensable place that oxen occupied in the farm economy of the pioneer era. Their number in this state was first counted in the New York State census for 1855, by which date they had already passed their zenith and were definitely on the way out. Nonetheless, there were reported more than 144,000 "working oxen," or more than 72,000 "yoke." Every county of the state except New York had oxen. King's County (Brooklyn) had forty-one. Even Richmond (Staten Island), now so completely urbanized, had 400. The premier county of the state was Dutchess, with 6,263, but the most noteworthy county was Sullivan where were counted 4,265 working oxen as compared with only 3,092 horses. As a matter of fact, it took Sullivan County a long time to get over the ox habit. I remember being in Monticello, the county seat, in the early years of this century when three or four yoke of oxen standing by the curb was still an entirely commonplace sight on Main Street. Today, a yoke of cattle in any village of the state would attract at least as much attention as an elephant.

In the era when oxen were in their heyday, none of the present popular breeds of dairy cattle had been brought to America. The most commonly found strain of cattle was somewhat of the Shorthorn type, colored red, red and white, white, roan, or brindle, with an infinite number of variations.

In 1800 some Devon cattle were imported to Massachusetts, and in 1807 representatives of the breed were taken to Otsego County. These Devons quickly won an especial reputation for their excellence as oxen. The breed has certain characteristics which were a basis for this favorable opinion. To begin with, every Devon was like every other in color—a solid light red— so that pairs were always matched, something that was a matter of pride on the part of many owners. Then, too, Devon oxen carried beautiful, spreading white horns which surely added to their impressiveness. As men phrased it, "Devons had style." But more important than these nonessential characteristics, was the fact that, typically, the breed had a certain snappy activity and light-footedness which gave them a special place in the ox world. Between 1825 and 1850 there were a number of importations, and there is no doubt that their reputation as oxen was the basis of the onetime wide popularity of the breed—an acclaim now long forgotten.

Properly to educate a pair of oxen for their life work was a job calling for a good deal of patience and skill, and it was best that their training be started in calfhood. Most farms, at least most farms where small boys were growing up, had a miniature ox yoke and in the absence of the organized and supervised recreational activities of these days, training calves to drive before a light sleigh or cart was a sport rather than a task. Probably few farm boys of a century ago failed to have this as part of their farm curriculum. By the time these calves were a year and a half old they began to play some useful part in the farm economy. In the

Training the Calves

everyday speech of the farm, young teams were always referred to as steers. The order was, steer calves, yearling steers, two-year-olds, three-year-olds. Only when they were four years old, practically mature, would the old-timer have called them oxen. Probably it should be said that in the speech of our local countryside we were inclined to use the term "cattle" rather than "oxen." If our man John Schaeffer had been reporting the doings of our neighbor, he would have said, "There were three yoke of cattle plowing in the twelve-acre lot."

I suppose there were certain eccentric or misguided owners

who gave their teams unusual or fanciful names, such as Star or Brindle—at least so I have been told—but the most common names for oxen were just two, Buck and Bright. It was as if all the dogs in the world should be called either Sport or Rover. This usage was surely almost universal over New York and New England, and perhaps farther afield. As one stood behind the pair, the ox on the left was always the nigh ox and his mate the off ox. The nigh ox was called Buck, while his mate was Bright, and this custom was so set that it had (to quote a phrase of legal jargon) almost "the force and effect of statute law."

Prideful owners embellished their teams by tipping their horns with ornamental brass knobs of various designs. In driving oxen, the gad, or whip, was an almost indispensable accessory. This was a four- or five-foot length of small hardwood sapling, with two or three feet of leather thong tied on the end for a lash. This whip was used mainly as an instrument of guidance and control rather than punishment. The words of command were few and about the same as those used for horses. There was "geddup" and "whoa" and "back," and, in addition, "haw, haw," which meant turn to the left, and "gee, gee" which indicated a right turn, these instructions being reinforced by appropriate flickings with the whip. In plowing, the job in which oxen excelled, certain intelligent and docile teams could be controlled wholly by the voice without the use of the gad. In the days when oxen were still common on our farms, there were occasional farmers who drove them before the mowing machine.

We had on this farm for more than fifty years a farm helper, one John P. Schaeffer. His advent here considerably preceded my own, and he was with us for a long generation thereafter. He knew both horses and oxen, but preferred the horses. I remember his declaration relative to driving: "I'd rather have the lines [reins] in my fingers than in my teeth." However, he did recognize merit in oxen. When in a reminiscent mood he used to recall a certain yoke of cattle we once owned that were remarkable for their ambition and lively gait, and especially for the fact that they definitely wanted to keep up with the plow ahead. In the graphic speech of the farm, his tale ran like this: "I used to plow ahead with Win and Fan, and Spence Mungo would follow me with the cattle, and old Bright would have his nose right in the seat of my pants all day." That was indeed high praise, because Win and Fan were recognized as better than average horses.

One more of John's memories: oxen when pressed without rest in hot weather would pant and show signs of distress. Humane and prudent drivers would stop and let them stand to "cool out." As John told it, we had an ox who eventually became so wise that in cold winter weather one had merely to go into the stable and rattle the iron rings on the yoke and immediately old Buck would begin to pant and display every symptom of exhaustion. Apparently he was not as "dumb" as generally supposed.

Another item of ox behavior (this from my father): oxen will never drink when overheated, whereas a horse under similar circumstances must be restrained from taking all the water he wants.

While the popular conception holds oxen to be patient and faithful, if rather stupid, brutes, yet if the truth be told they had their fair share of meanness and trickery. There were bad-tempered oxen that would kick, and to this day the expression "kick like a steer" persists as one of the most common phrases of our everyday farm speech. Also, there were some oxen that would balk, and even now a balky ox remains the symbol of unmovable stubbornness. Then, too, there were oxen that would run away, and, inasmuch as the reins were in the driver's mouth and not his hands, about all he could do was to holler "whoa, whoa," which didn't amount to much. Three thousand pounds of bovine bodies in a plunging gallop before an oxcart must have been a fearsome spectacle, especially if (as in one tale I remember) their course included a hop yard with standing hop poles.

Finally, there were certain tricky pairs that learned to co-operate in a veritable sleight-of-hand performance known as "turning the yoke." Every old-time ox driver knew something of this, and my father often talked about it. Turning the yoke was accomplished essentially in this fashion: the team, held closely together in front by the yoke, would separate as far as possible in the rear, throwing their butts out and away from each other until they were almost facing. Then one ox would lower his head while the other would raise his head and neck and his end of the yoke and pass it over the neck of his mate. When this fairly complicated maneuver was completed, all within a very few seconds, the pair stood transposed—the nigh ox on the off side and the yoke on upside down. Then the driver had the job of untangling the mess. The whole thing sounds like an invented yarn, but there is abundant testimony that certain pairs had full command of this trick.

It was always a question for debate as to whether oxen or horses could pull the heavier load. Usually a yoke of cattle had the advantage of greater weight than the horse teams of that era. This much is certain: among stumps and stones the oxen, with their freedom from eveners and whiffletrees, had a handiness that horses could not hope to equal. The single light chain running from the yoke to the load was an ideally simple tackle. It was in the rough, hilly sections that oxen were found in the greatest numbers and lingered longest.

In the days when oxen had a very large place in the farm economy, they were commonly shod after the manner of horses, because this enabled them to stand on ice. Moreover, if they were worked constantly on rough ground, their hooves wore down so fast that they would get tender-footed unless they were shod. An ox wore two shoes on a foot, one for each "claw." As a general rule, cattle refused to allow their feet to be picked up and held as the smith did when shoeing a horse. Hence some blacksmith shops were equipped with a cumbersome appliance known as "ox stocks." This was a strong chute into which the animal was led. All four feet were then firmly strapped to iron rings, a broad surcingle was passed under the belly, and by means of a block and tackle the victim was lifted clear of the floor. When he was rigidly bound and entirely helpless the shoes were nailed in place. It must have been a good deal of a job, but fortunately the hoof of the bovine is much tougher and so holds a shoe longer than the more brittle foot of the horse.

Oxen played a large part in the early days of our farm. My father's grandfather (on the distaff side), one Jared Goodyear, was a Connecticut Yankee brought up in a day and in a region when cattle were the most usual beasts of burden. In 1790 he and his twenty-year-old bride made their wedding journey in an oxcart, a journey that began at North Haven, and ended in Cayuga County in west-central New York. Ten years later, still with oxen, they backtracked as far as Schoharie County, where the family struck permanent root. All his life this pioneer had trusted in yoked cattle, and his sons carried on with the same faith.

In my father's youth there were sometimes three yoke on the farm. As a small boy growing up in the 1870's, I saw just the final years of that era; they are now only dim, hazy memories. In 1895 we built a new barn. There had been no work cattle on the farm for a good many years, but as my father contemplated the big job his traditional training reasserted itself, and he felt that he could not undertake it without a yoke of oxen. At that date oxen were becoming somewhat uncommon, but Henry Kimmey, who lived five miles to the north, sold him a pair for the occasion. That summer I learned to yell "Buck" and "Bright," and "geddup" and "haw" and "gee," and, in a word, to drive them very unskillfully. That brief summer remains my only firsthand experience with oxen. I believe my father was rather disappointed in this purchase. At any rate, after only a few months, he sold the oxen. I do not remember if it was to another owner or to the butcher. This yoke was the last representative of a long, long line, and without a doubt there will be no successors.

There are sound reasons for the large place that oxen occupied in pioneer America. For one thing, their first cost was low in an era when almost everybody was poor. Bull calves were a sort of by-product of cow keeping, and almost anybody could get hold of a couple of calves. Another point is that in a day when there was little land under the plow and grain was scant, the ox could do more work on hay and pasture than the horse. The ox team had no rivals as hay burners for motive power. Then, too, in working land cumbered with stumps, roots, and stones, the patient and supposedly dull-witted ox was better than the horse. Oxen almost never injured themselves. In deep snowbanks they would wallow serenely along under circumstances where a team of horses might rear and plunge and quite possibly cut their own or their teammates' feet with their sharp shoe calks. Also, the speedier horse is heir to a considerable number of diseases and disabilities which do not afflict the bovine. Finally, when a horse comes to the end of his course he is almost without value, but at the last the ox furnished a heavy carcass of what, in that day at least, was deemed acceptable beef. The pioneer could work his team as long as seemed wise, and then he could eat them after they were dead. Yes, in diverse ways the patient, and in some respects majestic, ox team played a tremendous part during two full centuries of our history. The ax and the ox-team together will always symbolize the conquest of the wilderness of eastern America.

The ox yoke was a wonderfully simple contrivance, and making it was easily within the skill of a good worker in wood. The local blacksmith forged the small amount of iron hardware that entered into its construction. In the later years of the ox era, a sort of standard, perhaps almost a statutory, price for an ox yoke was five dollars. Testimony that the yoke was usually of individual or community manufacture is found in the New York State census for 1855. It lists some 374 crafts and callings but it is not recorded that any man proclaimed his business to be making ox yokes. Nonetheless, there must have been a time when they were turned out in great numbers. An incidental advantage of oxen is that leather harness for horses is relatively expensive and grows old with the years, whereas an ox yoke is low in first cost and usually endures for the life of the owner.

Making an Ox Yoke

In 1812 at Pittsfield, Massachusetts, that astonishing, crusading Yankee, Elkanah Watson, promoted what is proclaimed as the first agricultural fair ever to be held in America. Among the amazing spectacles it offered was a line of sixty-two yoke of oxen all hitched to one plow and that plow held by the oldest man in the community. Given a man with Watson's enthusiasm and genius for organization, it would not have been difficult to have duplicated this show almost anywhere in the Northeast. Elkanah's teams and millions of others have marched down the road of the years and out of sight so that we may never see them any more unless we shut our eyes; the stentorian cries of "haw, haw" and "gee, gee" which once resounded on almost every farmstead have fallen into unanswering silence.

CHAPTER VI

The Log Home
in the Clearing

THE first European colonists evolved a distinctly new type of architecture. They had come from an old land whose forest resources had been so long depleted that wood, so far as the outside walls and roofs are concerned, was rarely used as a building material. Typically common folk lived in houses with walls of rough stone and roofs of thatch. The less common folk had dwellings of hewn stone or brick with roofs of slate or tile. But in the new land to which they had come, they found the choicest woods in great abundance. These firstcomers would have been singularly lacking in initiative and inventive instinct if they had failed to devise some plan for converting the forest into a home. The revered Pilgrim fathers left surprisingly detailed accounts of their social and political beginnings, but there are extant no blueprints of their architecture. It is unlikely that at the first they hit upon the precise type of building which a little later became the almost standardized construction of the frontier. It has been suggested that the first shelters at Plymouth were really half-dugouts with the roof and the front made from poles covered with bark.

There is abroad an idea, which has won considerable circulation and credence, that the Swedes are to be credited with the introduction of the log house and that from them the usage spread until it became universal. To me this tale seems apocryphal. Doubtless it is true that the Swedes, living in a country of unusual forest resources, were builders of log houses—a type of construction unknown in most of western Europe. On the other hand, the Swedes—the few who came—did not arrive until 1638, and never had anything more than a very limited and precarious hold in Delaware, eastern Pennsylvania, and southern New Jersey. It seems impossible that their isolated example could have set the architectural pattern for the whole Atlantic Coast and the hinterland. I believe that the log house was an invention evolved out of necessity and that it proved to have so many advantages and was so suited to the time and place that it won almost universal adoption.

Both tradition and the examination of the few authentic log houses still surviving in the Northeast agree as to their general architectural features. Typically, they were small, low, rectangular structures. Sixteen by twenty-four feet was the average size. Most commonly, the lower story was one room with a big stone fireplace, which served as a central heating plant. This ground-floor apartment was living room, dining room, nursery, and master bedroom as well. Above was a loft, reached by a ladder set against the wall, and here the older children slept; on occasions it served as a guest chamber as well.

Many of the log cabins had only dirt floors. Some of the pioneers, however, laid wooden floors. These puncheon floors were made of riven slabs of wood laid as nearly level as possible. The boards were then smoothed down with an adz. There were certain free-splitting trees—best of all the big, old, white pine— which could be riven into planklike slabs that needed only a little smoothing. The log construction did away with the studding and plastering which we think of as a necessary part of a house.

The roof, supported by pole rafters, was, in the beginning, sheets of bark. Probably the best bark for making roofs came from basswood, elm, and ash. Either pine or hemlock, if the trees were young and the bark not too thick, furnished satisfactory roof covering. Since most trees slip their bark very easily during the summer it was not difficult to obtain a sufficient amount of bark for a roof. The bark sheets could be laid into a roof that would remain tight for several years. Later, as the urgency for extreme haste had passed or as the family fortunes improved, the bark could be replaced with shakes—big, boardlike shingles riven from free-splitting timber. Not infrequently, in after years, the owner would wish for additional space and convenience, and, because sawmills and lumber were now available, he compromised by grafting a new frame addition to the old log structure which had served the family in earlier years. Also, he sometimes covered the logs with new siding, and when this was painted, the casual passerby would not notice that it was anything other than the conventional frame house.

There still remain enough authentic log houses scattered over the later-occupied regions of the Northeast to establish the fact that there were currently in use two fairly distinct types of construction. The more common, and surely the simpler and easier, method was to cut smooth logs of the required length with a diameter of ten or twelve inches on each end. Almost any kind of wood could be used, but white pine was the best because it was soft, easy to work, light in weight, and resistant to decay, and more often than most species it furnished logs that were smooth and of uniform diameter.

First, the logs were peeled, and then the builder, with sure, skilled strokes of his ax, cut a notch or "saddle" near each end of the log. If this notch was just half as deep as the diameter of the log, and if its companion log had a notch of the same depth and width and slopes of the same angle, it formed a rude right-angle dovetailing so that they might be laid up securely in cobhouse fashion. If the logs are approximately one foot in diameter, about fifty logs of various lengths will suffice to form the four walls of a house with one room seven feet high and with a sleeping loft above. Of course, even if erected by a skilled workman, a wall of logs was far from tight. To plug the larger openings the builder used lengths of wood split in triangular shape. The remaining crevices were sealed both inside and out with a puddled clay or, if time, funds, and material were available, with mason's mortar of burned lime. In the more elegant homes, the interior was whitewashed.

Such essentially was the log house in its most primitive form. Such scanty commentary as we have indicates that such a shelter could be "rolled up" with astonishing quickness. Dr. U. P. Hedrick, in his *A History of Agriculture in the State of New York*, estimates that a "log house with two rooms below and two above could be built by hired labor for $100.00." Considering the exceedingly low wages of that period, the primitive character of the construction, and the free raw material at hand, this figure is probably too high.

If the settler was a skilled axman he could, if necessary, cut his logs, haul them to the selected site, and notch them without any help. When, however, he came to the actual building he needed assistance, plenty of it. At least four good men would be needed, and a dozen could find employment. To assemble this help was easier than might be expected, because mutual assistance and neighborly co-operation was a firmly established code of the new country. When a new house was to be "rolled up," there was no labor shortage. Within an astonishingly short time following his arrival, the newcomer had four walls about him and a roof above. The big stone fireplace with its chimney, however, must have constituted a major problem and a strain upon his resources. This then was the typical and most primitive type of log house

Starting the Log House

built on the advance picket line of the invading pioneer. It was a definite architectural style which may properly be designated as primitive American.

While the foregoing may be deemed the typical log house, there was another type of construction used. I would not attempt to guess as to its relative frequency. For this second method of building, the logs were hewn to timbers of perhaps ten inches or twelve inches square. These were "halved-in" at the ends to make the corners of the walls (see figure). Once the timbers were thus prepared, the house could be put up very rapidly. When finished it had a neatness and smoothness not possible in a structure of round, notched logs. It was a log house de luxe, or rather, a timber house de luxe. This type of construction called for a great deal more labor—perhaps three or four times as much as was involved in the other type of house. In building the log house, each log had only to be notched, whereas in building the timber house each log had to be squared to definite dimensions with a broadax. It is doubtful that a timber house would ever have been attempted by any settler intent on getting some sort of shelter at the earliest possible moment. It is my theory that these relatively fine-hewn houses represented the second generation of houses, resulting when some settlers grew dissatisfied with their first rude homes and, having prospered in some small way, built a new and better house, albeit still of logs.

Log houses were often far from temporary structures. Many of them endured for at least a generation and some for a good deal longer. A case in point is the story of Thomas Chittenden. He was a tough-fibered Yankee who did his own thinking and set up his own standards. He was for twelve years president of the Republic of the Green Mountains, and afterwards governor of the newly established State of Vermont. Counting both dignities, he was for eighteen years chief executive of that tight little commonwealth. As the tale runs, some meddlesome uplifter once chided Chittenden because his simple dwelling so ill comported with his official position. The old gentleman testily answered that he had lived in his house for a good many years, that it suited him perfectly, and that if he was satisfied other people ought to be.

The log home has in eastern America a long and honorable history. During more than two full centuries, from the Atlantic seaboard to the grassy plains beyond the Appalachians, the hearth smoke of the log dwelling in the clearing marked the utmost west of the white man's march. As no other one thing, it symbolizes the American pioneer.

As late as 1855 we had more than 17,000 New York State families living in log houses. Within my memory, they were not too uncommon in the Adirondacks and in the remote valleys of the Catskills. In years as recent as 1890, one tiny Catskill valley, Cole Hollow, had three or four of these dwellings within a mile.

I believe there are in New York State a few, but very few, authentic old-time log houses still occupied as family homes. This does not include a certain number of modern pseudo-log houses with polished, hardwood floors, electric gadgets on every hand, and guest chambers with twin beds and tub and shower. These rustic retreats are now and again built by sophisticates who would like to make themselves believe that they are emulating their pioneer ancestors.

In the southern highlands—home of the mountain whites—there are still numerous log houses. Millions of Americans, some of them fated to become illustrious, have been born within walls of unhewn logs. Life for them lacked many of the comforts and refinements we of this softer age have come to deem indispensable, but, to some of these folk at least, their simple domiciles symbolized good childhood memories and "Home, Sweet Home." Of all the precepts of philosophy that have been offered to the world, there is none more vital than the ancient declaration: "A man's life consisteth not in the abundance of the things he possesseth" (Luke 12:15).

CHAPTER VII

The Story of a Farm

I TAKE considerable pleasure in the fact that I am removed from pioneering by so few generations. My mother's father was born in 1814 in the town of Carlisle, Schoharie County, in a log house. Along with a thronging brood of brothers and sisters he lived there until he was twenty-one years old. From the time I was a small boy until I was a man grown, grandfather was a member of our family so that I had always at hand authentic testimony concerning the ways of pioneer life. He always emphasized the snug warmth of a log house in winter. This is not to be wondered at because there was a foot of seasoned timber in the outer walls and this is more insulation than a modern construction is apt to provide. This built-in protection was particularly fortunate because the only heating system was the one big fireplace. Due to the need for maintaining fire for cooking and the necessity of keeping live embers overnight by burying them in the ashes, the great stone hearth was never cold.

Grandfather not only retained pioneer memories, he retained certain pioneer habits as well. His father had been a yeoman in

New England in a region where corn rather than wheat was the basic in the farm diet. The Carlisle farm to which he came proved better adapted to corn than wheat, and so it was that corn bread and various other culinary adaptations of corn were used instead of wheat bread. To the end of his long life he held to his youthful training and esteemed johnnycake in milk as one of the ultimates of gastronomic excellence. Our johnnycake was a glorified hoecake. Corn meal was the base; the other ingredients were milk, eggs, shortening, and molasses. It was baked in big pans. Another of grandfather's favorites was supawn, which was nothing more or less than corn-meal mush.

With the passing years the primitive living conditions on the frontier changed—often with astonishing rapidity. Within a single generation in the little villages and on the farms, log dwellings gave way to houses of frame and occasionally even of brick and stone. Some of these early houses were substantial and dignified. My grandfather's story is a case in point. As I have said, he was born and grew to manhood in a log house, but when he was twenty-one his family decided upon a more commodious and pretentious dwelling; so, on a pleasant knoll, great-grandfather built a two-and-a-half story red brick house, which is similar to those homes that lend such character and dignity to the Pennsylvania Dutch (German) countryside of Lancaster County. At the time it was considered one of the three or four best farmhouses in Schoharie County.

To move from a log home into such a dignified dwelling might indicate a great advance in the family fortunes; however, I do not for a moment believe that great-grandfather had accumulated a considerable store of money. Indeed, any such feat would have been impossible for a farmer who, only thirty years before, had come to the unmitigated wilderness with very little worldly gear, his main tangible assets being his strong young wife and his own vigorous body and stout heart. What it did mean was that in some small fashion he had prospered and was energetic and ambitious for better things. I think his fine new house was made possible because he had a brood of husky and industrious sons who were his co-laborers. The brick was burned on the spot, and the son who became my mother's father related how his particular task was to mold the brick. The clay for each brick was placed in a freshly sanded wooden mold. The sand kept the clay from adhering to the wood. Excepting the windows and a little hardware, the house was constructed of material produced within the family's own fence lines. Of course, there must have been hired masons and carpenters, but in that day wages were unbelievably low and a dollar was a large sum of money.

I should not neglect to add that great-grandfather did buy one pine tree. His farm occupied a hardwood ridge which grew very little pine, and pine was regarded as indispensable for interior construction. So he went to a farm only two or three miles south and purchased a single pine tree, paying for it the astonishing price of ten dollars. Doubtless, he got the pick of the woodland, and it must have been a veritable monarch of the forest because from it were split and shaved all the shingles for the house, and enough was left over to furnish the finishing lumber for the interior. This is the tale as it came to me from grandfather. It is interesting to note that, as early as the 1830's, choice pine lumber in eastern New York had a very substantial value, while a generation earlier it was burned without the least compunction.

The house thus completed served its builder and his children's children for considerably more than a century, and then, as homesteads frequently and sadly do, it passed into other hands. For a while the old house was sadly neglected, but it has changed hands again and it now has owners who are modernizing it and who cherish its traditions. If the roof is kept sound and if needed repairs are made now and then, another century or two may slip by without bringing visible evidence of decay or detracting from the simple, four-square dignity of the house.

Remembering the red brick house and the traditions of my

boyhood years, I am impressed that great-grandfather, John McNeill, lived and died serene in his faith that the pioneer age which had surrounded and nourished him would always remain the permanent pattern of the rural community. In the year 1835, when he replaced his log dwelling with a new brick house, that age was still dominant, However, a discerning mind could have noted indications of the impending doom of the homespun culture. Great-grandfather was obviously untroubled by any such forebodings. So it was that in the third-story garret of his new home he made special provisions for spinning and weaving, including a big reel for yarn built into the very structure of the house itself. This, to me, is incontestable evidence of his conviction that as farm life was then, even so it would continue through the future years. This, remember, was in the year 1835!

Nor did the next quarter of a century, which passed before he died, bring any particular change in his outlook or his faith. Great-grandfather had no considerable wealth. He had the 112 acres of land, which in his vigorous young manhood years he had painfully wrested from the forest, some livestock, and the few implements found on the farms of that period. Also, he had his hard-won savings, a few thousand dollars, a considerable sum in a rural community. I suppose that by himself and by his neighbors great-grandfather was reckoned a forehanded and successful man.

So it was that one November day in the year 1859, only a few months before his death, he decided that the time had come to answer the question, "Whose then shall these things be?" Doubtless, he sought the help of the village counselor-at-law because the introductory paragraph of his will was so rich in the jargon in which the law has always delighted:

In the name of God, Amen, I—John McNeill of the Town of Carlisle in the County of Schoharie and State of New York of the age of seventy-three years and being of sound mind and memory, do make and declare this my last will and testament in the manner following, that is to say . . .

This rolling period having been indited, John McNeill proceeded to the business of the occasion. During the years there had been born to him and to Alice, his wife, a brood of four sons and three daughters. Two of his daughters had married, happily, as the years proved, but there remained one unchosen girl who, it seemed, might expect no other career than that of a daughter in her father's house. It was a day when it was held that a son might fend for himself and a daughter who had found a mate depend upon him for all things she might need, but the daughter who unfortunately had somehow missed the gate of marriage was properly entitled to special consideration, care, and tenderness. So, having all these things in mind, he added to the counselor's introductory paragraph, these provisions:

First: I give and bequeath to my wife, Alice McNeill and my daughter Sarah Ann the east one-half of my house situate in Carlisle, Schoharie County, and use of the hall to pass to and from the same at all times during the time they shall occupy the said house, all of which they are to have and occupy themselves during their natural lives, and two good cows the same to be kept on the farm, and all my household furniture and seasoned wood to be fitted for such stove or stoves as they may have and to be delivered in wood box by said stoves; and five pounds of wool yearly during their or either of their lives and to be carded ready for spinning. And to be carried to and from church and such other place or places as they may wish, and they must be provided with food and clothing in good farmer's style in sickness as well as in health. The above request to my wife Alice to be accepted and received by her in lieu of dower.

To my son, Barzillai McNeill, the One Thousand Dollars I have paid to and for him in the purchase of the farm he now occupies. To my son, Squire McNeill, Two Hundred Dollars to be paid within two years after my decease. To my son, James McNeill, a promissory note I hold against him, amount One Hundred and Ninety Dollars besides interest bearing date March twentieth, One Thousand Eight Hundred and Forty-nine.

To my son, Henry McNeill, Two Hundred Dollars to be paid within four years after my decease. To my daughter Lucy Jane Matchin, One Hundred Dollars, to be paid one year after my decease. To my daughter Merry Shafer, if she outlives me, Fifty Dollars to be paid within five years after my decease.

But when he had called each of his children by name and had given to each their pitifully tiny inheritance, he bethought himself, and for the second time turned to the problem of his unwed child and made for her a further provisional remembrance: "And to my daughter Sarah Ann aforesaid if she gets married, Two Hundred Dollars to be paid within one year of my decease, and one good cow at the time of her marriage."

As it turned out, her father's hope that Sarah Ann might some day marry never was realized. She lived to be an old, old woman, continuing to the last to occupy her east half of the red brick house. In my boyhood I knew her well because she was my Aunt Sarah, a sister of my mother's father. I suppose life offered her little of privilege or opportunity, but in her humble fashion she was a good woman, zealous for good works and the Presbyterian Church.

Once a year when my grandfather was still alive, she would come for a fairly lengthy visit. Immediately on settling down she would demand some task to which she could set her hand. Her love of mending and darning was hardly less than a passion. She mended grain bags and the family buffalo robe and old carpets. She darned the accumulation of holey stockings, worn woolen mittens, and anything else that was darnable. She knitted gray stockings and blue-and-white striped stockings, and once or twice even insisted on getting the spinning wheel out of the garret where it had reposed from Civil War days. I suppose she was more exemplary of the homespun age than any person I ever knew. To her it was not a tradition nor yet a memory, but a very definite way of life.

I believe that my great-grandfather, having made such liberal and far-reaching provisions for his widow and for his girl, felt that the important business of the occasion was finished. However, there were certain formal details to be added:

Aunt Sarah's Visit

I give and devise to my son Merritt McNeill his heirs and assigns all that tract or parcel of land situate in Carlisle, Schoharie County, and State of New York, and known as the Homestead Farm and containing one-hundred-and-twelve acres of land, bounded as follows:

West by lands of James Sweetman, North by George Taylor's lands. East by the Highway. South by Roswell Huntington and George Vanderwerker's lands: Together with all the hereditaments and appurtenances thereunto belonging or in any wise appertaining.

To have and to hold the premises above described to the said Merritt McNeill his heirs and assigns forever provided nevertheless he pay or cause to be paid the several legacies or sums of money as I have directed and ordered to be paid.

I give and bequeath all the rest, residue and remainder of my personal Estate goods and chattels of what nature or kind soever to my said son Merritt. And lastly I do hereby nominate and appoint my friend David S. Brown of Carlisle to be the Executor of this my last will and testament hereby revoking all former wills by me made.

In witness whereof I have hereunto set my hand and seal this twenty-ninth day of November, One Thousand Eight Hundred and Fifty-nine.

JOHN McNEILL   L.S.

There are two respects in which this will constitutes a commentary upon the homespun age. First, it is evident that Yeoman McNeill, on the eve of the Civil War (five days before John Brown was hanged at Charlestown, Virginia), was entirely confident that the social and economic conditions then prevailing would remain the order of life through all the coming years.

And the other is that it reveals how very little cash money circulated among farm people at this period. Such small wealth as existed was expressed first in land and then in such tangible things as cattle and farm gear. The dollars he could distribute were quite insignificant in amount, and yet he lived in what was at that date the best house in the Township of Carlisle, and he was regarded as a man of substance and standing in the community.

In however kindly fashion we may try to view or even romanticize the pioneer era, we cannot escape the fact that it was a day of relentless and almost unmitigated toil. Without, the man was confronted with the herculean task of clearing the forests, subduing the land, fencing the fields, and sowing and harvesting the crops aided by only the most primitive of agricultural implements. Within her home, the woman struggled with the task not only of feeding her family but of clothing them as well. Retting, breaking, and hatcheling the flax, and carding, spinning, and weaving the wool would be to us today an endless, a wholly impossible undertaking. Yet frequently we find her in the field binding the sheaves as the grain fell from her husband's cradle. No wonder that the dearest boon the pioneer could imagine was rest—just physical rest. This great longing is expressed in their hymns. Often in bleak frontier churches the congregation lifted up their voices and sang:

> There is rest for the weary
> There is rest for the weary
> There is rest for the weary
> There is rest for me.
>
> .　.　.　.　.　.
>
> On the other side of Jordan
> In the sweet fields of Eden
> Where the Tree of Life is blooming
> There is rest.

Of course, not all pioneers were churchmen. Far, far from it. Nevertheless, it is certain that the church came in almost with the first wave of settlement and usually in advance of any organized school. Religion has always played a large part in the life of the typical rural community. The people were weary and heavy-laden, but to some of them more vividly than to us, I think, it was given to see sunshine serene and eternal on the Other Side.

CHAPTER VIII

What Crops Did the Pioneer Grow?

IN A world whose outstanding characteristic seems to be a bewildering rapidity of change and a constant submergence of the old by the new, it is surprising to note that the crops grown by the early American pioneer were essentially the same crops that dominate the northeastern states today.

Wheat, rye, barley, and oats were the four cereals of Europe and more distant parts of the world. They have been known and harvested for centuries, and their origin is so buried in the past that their natural habitat and, to some extent, their botanical relationships are a matter of conjecture and debate. About all we may say with certainty is that they were of European or Asiatic origin and (except possibly in the case of the oat) have no close New World relatives. They were so indispensable and so universal in the established farm economy that doubtless the seeds of all four were brought along on all colonizing ventures to America. Two other plants, peas and buckwheat, eventually came to have an important place in pioneer agriculture.

Wheat, of course, is par excellence the bread grain of the world, and, in those fortunate regions where soil and climate permit its production, it is the well-nigh universal loaf. The wheat flour of the pioneer was not so highly refined but far more nutritious than much of the wheat flour we are familiar with today. Now, unhappily, wheat is rather demanding in its soil requirements. On soils that are of low fertility, sandy, or poorly drained, it yields only precarious and scanty crops. Rye is a much hardier plant and will give fair yields on soils and under climatic conditions where wheat will fail. So it is that in much of northern Europe it is a loaf of rye rather than wheat that is the staff of life. Such traditions and statistics as are available indicate that rye was used considerably in the diet of the homespun age. The protein of rye is somewhat like the gluten of wheat in character so that to some degree it may be leavened with yeast, but it cannot, at best, attain any such degree of lightness as bread made from wheat. Rye bread was dark, hard, and tough and most people preferred the flavor of wheat bread. As a matter of fact, the rye bread in our bake shops of today is not straight rye but a mixture of wheat and rye flour, probably with rye the minor component.

The picture of the bread-grain situation in the State of New York in 1845, which is the earliest date at which statistics are available, and, which may be reckoned as a period still within our purview, is as follows. The total population of the state was over 2.6 millions, of whom more than 370,000 lived on Manhattan Island. Already, there was a really sizable city at the mouth of the Hudson. In 1845, the state grew 13.4 million bushels of wheat or just about the requirements of her total population if wheat were to be eaten universally. Everybody wished to grow wheat, and every county in the state grew some, even including New York County (Manhattan Island), which reported four acres with a yield of fifteen bushels per acre. Incidentally, that year the island had produced more than 6,000 bushels of corn, which indicates that there were considerable stretches of farm country in the hinterland north of Canal Street. Far to the north and west was the "Genesee Country"—a generic term including the Finger Lakes region and the Ontario shore. This was the largest area of high-class wheat land east of the Alleghenies, and already the recently come settlers were growing wheat in surprising amounts

Grain Barge on the Erie Canal

and floating it to tidewater via the Erie Canal and the Hudson River.

From New York much of the wheat was shipped to Europe. That era is still recognized as a very happy period in the agriculture of that region. Presumably, in this fortunate land everybody ate wheat bread and scorned all inferior cereals. The first wheat county of the state was Monroe, with a production of nearly 1.4 million bushels. Monroe was the only county of the state exceeding the million-bushel mark. Livingston County, however, had a larger per capita production: more than twenty-five bushels for every inhabitant. The amount of wheat produced was more than five times the bread requirements of the county.

This astonishingly intensive wheat production in certain regions of the state did not alter the fact that there were many localities that did not then grow, and probably have never at any time grown, wheat enough for local consumption. A striking case in point would be Greene County, which in 1845 grew only about one peck of wheat per capita or less than a three weeks' supply. Delaware is a county noteworthy agriculturally for its wonderful pastures and dairy herds. However, it has a soil ill-adapted to wheat and at this date she grew less than one-third enough wheat to supply her own population.

There were two possible solutions for this bread shortage. Conceivably, the pioneer might have settled the question by purchasing wheat or flour from some more fortunate region to fill out his own insufficient supply. It is doubtful if this was ever done in any general way. To begin with, since it was an age of almost universal poverty, the pioneer had always a struggle to come by the small sums of cash that he absolutely needed. Save in rare instances, his financial resources would not have allowed him to purchase his own bread. Then too, the idea would have been contrary to the whole philosophy of the frontier. The most firmly held canon of the pioneer was this: that the farm must be self-sufficient and self-contained, and, to the last degree possible, the family must live within the farm fence lines. So it was that he made a virtue of necessity and supplemented his insufficient supply of wheat by adding to his daily menu the cereals rye and corn and the noncereal buckwheat. A substitute for wheat bread was found in "rye-and-Injun," a type of bread made from rye meal or flour and corn meal. Rye bread has a very stiff crust and a particularly durable interior, while corn bread is always crumbly and easily broken. The two in proper combination may have counteracted and mollified each other and resulted in a product superior to either alone. This is only undocumented surmise. At any rate, rye-and-Injun was once a staple loaf in many localities, and the name and the tradition, if not the actual presence, have lingered down until a time within the memory of living men.

In passing, we should not forget buckwheat. All of our grains, including rice and the millets, the so-called cereals, belong to the great botanical family of grasses, a group with which buckwheat has no remote relationship. Botanists believe it was a plant native to Asia, whence it was introduced into Europe and cultivated there for many centuries. Some of the first colonists brought buckwheat to America, and it promptly became an accepted and indispensable part of the farm economy of the Northeast. It was once appropriately called beech wheat from its striking resemblance in shape to a beechnut. In 1845 it was grown in every county and in practically every township of New York State. Even Manhattan Island had a crop which yielded three hundred bushels. The total state production was 3.6 million bushels, about one-quarter of the entire yield of wheat and substantially greater than the crop of rye.

Buckwheat's large kernels easily mill into a fine, white, but rather gritty-textured flour. It is said that in Europe it has been used in various bread mixtures, but in our American practice its use as food has apparently been confined to pancakes. The orthodox procedure for pancake manufacture is this: pure buckwheat flour is made into a batter with sour milk or, much better, with old-fashioned buttermilk. This batter is made at night and put into a very large pitcher. To it is added a little of the yeasty batter left over from the morning baking. The pitcher is set in a pan and kept in the warm corner behind the kitchen stove. And because during the night the batter will rise and, very frequently, run over, at breakfast time the foaming mass must be stirred down. Then the pancakes are poured on a properly heated griddle. Lo! beneath the baker's gaze, they blossom into brown, tender, fragrant discs, which have often been acclaimed as the world's highest achievement in this class of culinary arts. It may be that sometime you will meet a man who will claim that he was born on a farm in the buckwheat belt more than a half-century ago. You may examine his pretensions after this fashion: cross-examine him relative to his boyhood knowledge of buckwheat pancakes.

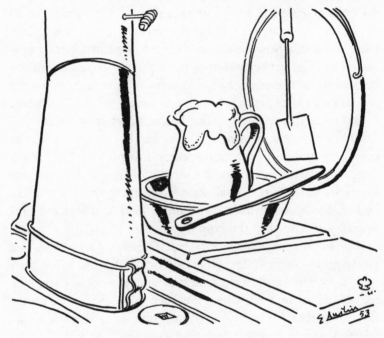

Batter for Buckwheat Cakes

If he exhibits familiarity with and steadfast loyalty to the institution, you may accept him at face value. If he betrays any unfamiliarity or falters in his speech, you may dismiss his claims and denounce him as an impostor. At any rate, it is certain he is no farm boy of the last century. Still, old customs and folkways linger long.

I know of only one drawback to this paean of praise. There is or was an opinion amounting to a conviction that there were

certain unfortunates who by sad mischance were allergic to buckwheat so that if they indulged in this royal food to excess they
developed a skin eruption, inelegantly but correctly designated
as the "buckwheat itch." I cannot say how far this was the truth
or in how much it may have been a diagnostic error. I only know
there are plenty of oldsters who would deny that the phrase
"in excess" had any meaning when applied to the laudable habit
of riotous consumption of buckwheat "pannie-cakes."

The early colonists brought one other grain from the European
homeland: peas. This was not the familiar wrinkled green pea of
today. It was the hard, round, smooth, gray pea commonly called
the Canada field pea. This type was not ordinarily used as food.
It was widely used as feed for stock, for swine. In that earliest
year for which we have statistics, we grew more than 117,000
acres in New York State. The yield exceeded 1.7 million bushels.
In 1845, peas were grown in every county of the state except
Manhattan Island.

The crop disappeared from our agriculture with remarkable
abruptness. In my father's boyhood, there were twenty acres of
peas for grain on our farm, and yet I cannot remember even a
single acre. The 117,000 acres of 1845 fell to only 48,000 acres
in 1855, and by the 1870's the crop was practically extinct in
New York State. Jefferson was always the premier county of
the state in pea production, and there an occasional field could
be found until comparatively recent years.

Two distinct reasons account for their disappearance. Harvesting peas was always a slow, laborious task because they had to be
mown with the scythe, and, after the coming of the mowing machine and the reaper, farmers turned to grains that could be
harvested by those implements. Probably the major reason was
the infection of a soil disease called root rot, which made the production of peas a precarious venture. In any case, a crop which
a century ago had a very important place in our agriculture is
only a memory today.

I have never heard or read that either barley or oats was part of the diet of the American pioneer. Since wheat, rye, buckwheat, and corn were available and could be converted into flour in the crude grist mills of the age, there would seem to be no particular reason why the pioneers should experiment with these other grains. Also they had a new grain—a grain that their forefathers back in Europe had never known—corn (or Indian corn, as it was sometimes called, or maize, as the botanists have named it). Without question or debate Indian corn was the greatest contribution the New World made to the food resources of mankind.

As is the case with all plants that have been cultivated for many centuries, the original form and native home of corn are uncertain. The weight of botanical opinion supports the belief that it is a giant grass native to Mexico or more southern countries. At any rate, whatever its origin, when the first European colonists landed in America they found that corn was known and cultivated by all of the Indian tribes on the Atlantic seaboard and far into the interior. Under their untutored selection it had taken on a diversity of form and character. The newcome immigrants were familiar only with the grains of the homeland—wheat, rye, barley, and oats. It is surely a tribute to their open-mindedness and resourcefulness that from the very beginning they adopted the practices of their Indian neighbors and made corn the premier grain, not only in feed, but in food as well.

In New England the use of corn as food lingered long after the homespun age had drawn to a close. Something more than sixty years ago, when I was in the College of Agriculture at Cornell, F. G. Bates, my most intimate friend, was a student in the College of Arts. He was a Rhode Islander, born and bred in the orthodox traditions of the hill country Yankee farmer. At his home the almost fabulous Rhode Island johnnycake was the daily fare, the staff of life. In the wheaten bread regime of an Ithaca boarding house, he bemoaned its absence. He insisted that making good Rhode Island johnnycake was something of an occult mystery. In its construction there was no place for the cheap, yellow corn meal from the corn belt. Johnnycake that was worthy of the name must be made from home-grown, white flint corn ground under millstones by water power, but surely not by steam. Now doubtless, this insistence upon this particular home-grown corn will be regarded as merely the harmless exuberance of local pride and patriotism, but some chemist making analysis of corn from many sources stumbled upon the fact that this particular variety had a fat or oil content far above the average, and so the old notion was justified, and this corn was, in that day, really superior for human consumption. One year I grew a little plot of that special Rhode Island corn on Hillside Farm here in east-central New York. It proved to be a very distinct variety— dwarfish, much inclined to sucker, and having three or four little ears of hard, white flint corn on a stalk.

Take note that all this was sixty years ago in a different economic era, when among New England farmers ancient manners and customs still lingered. With the years these customs have disappeared. Doubtless, there remain few households where johnnycake is still daily fare. The thin and stony fields on which Caleb Bates grew his special corn are now planted to other, supposedly better, sorts of corn or are again engulfed by the expanding New England wilderness.

The Indian tribes of New York and New England grew flint rather than dent corn, and, inasmuch as the white man adopted corn growing from the Indian, it follows that the corn grown in the Northeast during the pioneer period was exclusively flint corn. Not until the coming of the silo made the larger varieties desirable, were the dent varieties grown to any extent in this New York–New England region.

On many New England farms and on some old-fashioned, conservative New York farms the variety known as King Philip may still be found. It is one of the oldest sorts with a name. Its red color, together with its long, slender, eight-row ears, make it particularly distinctive. It was a favorite with Yankee farmers almost from the beginning, and there is little doubt that it is named for that militant sachem of the Wampanoags. He was Metacomet to his tribesmen but the English knew him as Philip. In 1674 he organized a confederacy of southern New England Indians, who launched a valiant but vain attack against the growing power of the white man. The result was a lurid chapter of ambush and murder which terrorized all that region and is known now as King Philip's War.

The sweet corn of our gardens of today represents a very marked departure from either flint or dent corn as grown commercially. Sweet corn was not found among the seaboard tribes and was unknown to the colonists until about the time of the Revolution. There are at least two circumstantial accounts of its origin. One is that a certain Richard Begnall found it growing among the Indians of the lower Susquehanna Valley. A second is that soldiers in Sullivan's raid found it among the Senecas of western New York. Parts of the Susquehanna Valley constitute about the finest corn country to be found east of the Alleghenies, and the Senecas were, according to the standards of the red men, most noteworthy farmers. Either locality may easily have been the cradle of the distinctly new type of corn.

Corn is in a way a generic term which includes all four of the cereal grains, but in England the word commonly referred to wheat, while in Scotland it meant oats. Therefore, when the first colonists to land on our shores became acquainted with the new cereal it was the most natural thing in the world that they should distinguish it from the familiar grains of the homeland by the simple device of prefixing the word Indian, and the name stuck. Eventually, it became a firmly fixed term in our language. The term Indian meal was familiar to me in boyhood days. It partially survived in rye-and-Injun bread, and to this day on the most sophisticated menus we read "Indian pudding," so named because its basis is Indian meal.

Although census figures or exact statistics are lacking, it is certain that up until the close of the homespun age, which I have rather arbitrarily assumed ended about the period of the Civil War, it was maize rather than wheat that was the staple fare of farm folk. This statement would not be true of certain favored regions such as the Genesee country of western New York, or many fortunate valleys of New York and Pennsylvania where the soil is suited to wheat growing. There, almost from the beginning, farmers ate wheat bread to the full, and, possibly, were a little contemptuous of less lucky folk who must supplement their scanty wheat supplies with a good deal of corn. Nonetheless, barring these favored exceptions, throughout most of New England, much of New York, and most of the seaboard country south of Pennsylvania it was corn rather than wheat that constituted the major bread grain.

Often, through the years, artists have busied themselves making seals or insignia or coats-of-arms which would symbolize agriculture. The old English poet Cowley, writing in the heraldic terms of his days, says that "A Plow on a Field Arable Is the Most Honorable of Ancient Arms." Other men have used the plow as the central motif of their creations. Another favorite device has been a sheaf of wheat, often crossed with a sickle. These pictorial representations are all germane to agriculture, but as a symbol of agriculture in the pioneer period of northeastern America the best emblem would be a little field of corn shocks standing in orderly array with yellow pumpkins lying in between.

CHAPTER IX

The Crops of
the Homespun Age

THESE, then, were the grains of the northeastern states in pioneer times: wheat, rye, oats, barley, buckwheat, and peas. All of these are ancient crops whose beginnings are concurrent with the beginnings of civilization. To these was added the grain new to Europeans, maize or Indian corn.

Wheat, imperial bread grain of the world, has had a very special place in our agricultural thinking. While its peculiar desirability has always been recognized, there has never been enough wheat in the world for everyone to eat the wheaten loaf. So it was looked upon as almost wicked to use wheat for animal food. This was surely the case throughout the experience of the northeastern pioneer, and even in much later times. I can remember when the only wheat we ever thought of using for poultry feeding was the few bushels of wheat screenings—the shrunken and broken kernels screened out in cleaning the crop for market. To feed good wheat to animals would have been akin to a shameful, impious deed. The opening of vast new wheat

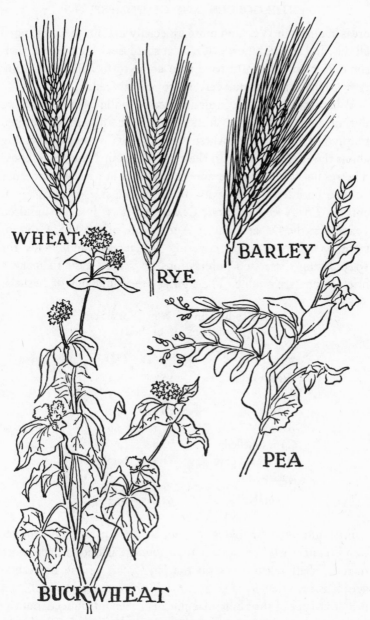

WHEAT

RYE

BARLEY

PEA

BUCKWHEAT

Crops of the Homespun Age

areas in our own West and more especially in Canada has changed all this, and at times the world wheat surpluses have been burdensome. Yet only recently have men come to feel that it is really right to make this unique cereal into animal feed.

What has taken place in grain production in New York State during the past hundred years? That year 1845 must be cited simply because it is the earliest date at which we know anything about the crop acreage and the total yields in New York State. The results of that enumeration are bound in a very ponderous volume (printed, by the way, on only one side of the paper), entitled *The New York State Census for 1845*. Just one hundred years later the Department of Agriculture of the U.S. Government issued its statistics covering the acreage and yields for 1945. Using these two publications, it is very easy to make a brief comparative table. The figures are in millions of bushels:

Grain Production in New York State
(in millions of bushels)

	1845	1945
Wheat	13.3	9.1
Rye	3.0	.2
Oats	26.3	19.2
Barley	3.1	2.6
Corn (shelled)	14.7	4.6
Dried field peas	1.8	Negligible
Buckwheat	3.6	1.5
Totals	65.8	37.2

First and unmistakable is the fact that so far as grain production is concerned our present total tonnage is only a little more than one-half what it was in the later years of the homespun age. It is true that in 1845 a crude form of reaper, as well as a primitive type of threshing machine, had been introduced, but the use of these machines was as yet exceedingly limited. The larger part, surely more than ninety per cent of that sixty-six million bushels of grain, was cut with cradles, bound in sheaves with straw bands, and threshed with a flail or under the feet of horses. Comprehending, as I do, something of the hand labor involved, I cannot but be appalled at the backbreaking toil that went into the production of that great quantity of grain. The estimate applies only to the harvesting, but before that was the plowing and the preparation of the ground by the use of poor walking plows and primitive harrows, in very many cases pulled by slow but patient oxen. If, as an earlier and sterner generation used to believe, relentless and uncomplaining toil be a cardinal virtue, then the farmer of the premachine era must be deemed its highest exponent.

The production of each of our grains is markedly less than it was a century ago, and peas and rye seem to be nearly extinct. New York State is fortunate in having considerable stretches of excellent wheat land where that grain has been grown from the very beginning. There is no reason why it should not continue to be a major crop even if less important than it once was. What is true of wheat is also true of oats and barley, although barley was never a really major crop in New York State. The pioneer housewife made bread from wheat and rye, and she devised different ways of using corn. She made numerous buckwheat pancakes, but oats and barley were apparently not recognized as proper food for the family. The production of corn to be husked for grain has greatly declined, but if we count corn grown for ensilage, the total yield would be more than any figure ever attained in the old days.

The production of rye has declined considerably and it seems on the way to virtual extinction in New York State. There are two reasons that explain its decline. One is that rye, while a hardy plant that may give fair returns under unfavorable conditions, does not equal the yield of wheat where soil and fertility conditions are suitable for that more exacting cereal. The other reason is that rye straw has long been recognized as the very best bedding for horses, and in the past the straw was in demand for that particular purpose. The straw was worth as much as the grain, but the passing of the horse has done away with that once profitable outlet.

Buckwheat, while grown in every part of New York State a century ago, was never important in a country-wide sense. In 1945, the total production of the entire United States was only 6.6 million bushels, of which almost exactly one-half came from Pennsylvania and New York. These two states still make up the buckwheat belt.

The little, hard, gray field, or Canada, pea was once an important crop in many counties of New York, but it is now extinct or so uncommon that the federal crop reporting agencies make no effort to estimate the acreage or yield. In 1945, the yield for the whole United States had fallen to a little more than half a million bushels, or less than one-third the yield in this state alone a century earlier.

Meanwhile, one—and only one—new grain has been introduced into our New York agriculture. Soy beans have been discussed and recommended for at least a third of a century; yet, in 1945, the total yield of dry beans was only 56,000 bushels—a production that may be called relatively insignificant. Somehow or other, this valuable new crop has never succeeded in really fitting into our agriculture.

Hops are a crop that, while never important in the sense of being fundamental to the farm economy, was widely grown in a limited way and, hence, deserves at least some words of remembrance. The plant is interesting botanically because it is found wild on both sides of the Atlantic, and botanists include these widely separated forms under a single species. It is uncommon for plants growing on different continents to be so closely akin.

Hops have been cultivated in Europe for at least a thousand

years, and they were brought to New England, New York, and Virginia soon after the first settlements. Doubtless their main use was in brewing, but they also found a place in the domestic economy. In my boyhood, the making of "emptin-cakes" was a regular occurrence in the kitchen. The basis of these cakes was a dough made of corn meal to which yeast had been added. The resulting fermentation was allowed to go on for a time until it was checked by drying. The resulting yeast plants would then lie dormant until given water, warmth, and starch. When used, an emptin cake was broken up in warm water. Once this was added to a batch of bread dough and put in a properly warm place, the whole mass would promptly rise. The making of emptin cakes, patting them into little cookies, and putting them on a bread board to dry was a regular part of household procedure.

For some reason, it was believed that the corn-meal dough had to be moistened with a decoction of hops. I am inclined to doubt if this was in any way essential for the desired result but it was the accepted ritual. Thus every farm family had, near the kitchen door, two or three hop vines. All this was common practice during my boyhood, but it has fallen into disuse for more than sixty years.

Hops had also a place in medicine of pioneer days. People drank hop tea and used poultices made from hops. It is probably true that the warm aroma of hops does have some soporific effect— at least I remember most precise and enthusiastic endorsement of the "hop pillow" as a means of overcoming sleeplessness. Later on, toward the close of the homespun age, when long-distance transportation became available, and the brewing industry increased, hop growing became a highly specialized business in about five counties of central New York. Its palmy days and its long decline have the makings of a glamorous story.

Now, be it observed, that because our grain production is only a little more than half as large as it was in the closing years of the homespun age is no reason for believing that New York farming is on the way out. What it means is that our agriculture has been vastly diversified and intensified. It is true that peas, flax, and root crops, all important a hundred years ago, have disappeared from our agriculture, but in their place have come certain new crops and, far more important economically, a vast commercialization of certain products which the pioneer never grew beyond his immediate family needs. Anything resembling our present-day floriculture or horticulture or canning factory operations or our commercial vegetable-growing program was simply nonexistent in that fabulous homespun era. Of course, even in the most primitive times, the farmer must have at least a little money. So, from the beginning, he tried to produce certain crops which could be sold, but, nonetheless, in the pioneer period farming was essentially a subsistence rather than a business enterprise.

CHAPTER X

Meadows and Pastures in Bygone Days

IT GOES without saying that any type farm on which there are animals must provide not only grain for the cattle but also pasture for their summer subsistence. Fortunately, we have a number of wild grasses which tend to grow whenever the forest is removed. Also, even well-fed cattle and sheep love to browse the leaves and young shoots of most deciduous trees. A field that is continuously used as a pasture never has any hedgerows. This ability of grazing animals to keep down and eventually entirely kill sprouts and low brush is nothing less than providential, because if they are turned into newly cutover land they accomplish the double purpose of nourishing themselves and subduing the land for cultivation. In fact, if it were not for the help of the farm animals, it would be almost impossible to keep down the stump-sprouts and underbrush which so quickly spring up in every forest clearing. So it was that the new settler could depend upon some pasture almost from the beginning, without sowing any grass seed. But getting hay enough to carry his animals through the winter was a tough problem.

Although our native grasses produce fairly good pasture, they are not tall-growing and furnish only a scanty hay crop. Bluegrass —the grass that made Kentucky famous—is indigenous to the northeastern United States. Where it is at home it is regarded as pasture grass par excellence, but its possibilities for hay are rather limited. Red clover and timothy, both introduced from Europe, were the main plants sown for hay before the Civil War. With the years, the list of forage crops has lengthened greatly, but these two old stand-bys are still the leading hay crop in the northeastern states.

White clover is almost invariably present in old pastures. It is an old European plant and doubtless was imported by chance in the hay that was brought along to feed the first cattle to reach America. In any case, white clover kept pace with the most advanced line of settlement. A tradition says the Indians noted that when the white man appeared so did the white clover. They concluded that it could grow only where the white man had trod and named it "The White Man's Foot." Another tale related how the honeybee, never known in the Western Hemisphere until brought from Europe by the early colonists, was recognized by the Indians as a new insect in their experience and hence they called it "The White Man's Stinging Fly."

Previous to the introduction of the fanning mill, which came in with the second quarter of the nineteenth century, clean, commercial grass seed, as we know it, was not available. The farmer depended upon native grasses and what pasture land he could develop by scattering hay chaff and the sweepings of his barn floor and hay mows on the land. It was a primitive and inefficient method, but there was no other way in the premachine era.

In the middle third of the past century, root crops (turnips, mangle beets, and carrots) were important in New York State agriculture. In 1845, there were more than 15,000 acres of turnips in the state. The beets and carrots are not enumerated. To some extent, this was a carry-over from European farm practice, as was the raising of the 6.5 million sheep in the state at this period. Good shepherds had long known that roots had special value in sheep husbandry. Root crops give high food yield per acre under favorable conditions, but the great amount of hand labor required to grow and harvest them has practically eliminated them in this mechanized age.

When the homespun age was in flower, flax could be found on nearly every well-ordered farm. Along with the ax and the log house, flax seems in a very special way to symbolize this era. The outstanding place that it occupied in the pioneer economy and the numerous handicraft operations through which it passed before it became homespun cloth entitle this crop to special consideration in a discussion of the housewifely arts.

Then there was the rather diverse list of crops which we speak of as garden vegetables. First in economic importance is the potato. Such historic references as we have indicate that the Atlantic seaboard had been occupied for a century before that now almost indispensable tuber was added to our diet. Botanists agree that the potato was of South American origin and came to North America via Europe. It is said to have been taken from Spain to England in 1586, but it was slow in gaining popular favor. By the last quarter of the 1700's, it was of prime importance in Ireland. The ghastly famine of 1846 came about because of the sudden and almost complete failure of the crop which had become the Irish staff of life.

Probably there is no possibility of determining just when potatoes were first produced in any amount by our northeastern farm people. There is an interesting tale concerning how potatoes were served as a rare, exotic delicacy at a Harvard College Installation Dinner in 1707. There is the more sober tradition that a company of Scotch Presbyterians who landed at Boston in 1718 brought with them the first potatoes ever seen in New England. Once they had become common in Europe, it could not have been long before they were brought to some seaport town in America.

While potatoes go back into pre-Revolutionary times, they do not, like some of our other garden vegetables, go back to the very beginnings of colonization. Once introduced, they were certain to be widely adopted because of the ease with which they could be grown, the large amount that could be produced per acre, and the variety of ways in which they could be prepared for the table. I think perhaps my own father's early recollections of the crop may be worth recording. The destructive potato blight was unknown in America until the late 1840's, while the potato bug or, more accurately, the Colorado beetle, did not appear until about 1876. My father could remember the time before either of these infestations. He testified that time was when potatoes were planted with very little care or attention to fertility and yet the yield was abundant and certain. Any potato grower knows how the picture has changed since those halcyon days. Now we have not only the blight—at least two kinds of blight—and the bug, but a long list of diseases to which the potato is subject. Farming in the old days was a matter of relentless toil and primitive methods, but it is good to know that in certain ways there were compensations.

Beets, turnips, carrots, parsnips, and cabbage are all ancient European plants which have been cultivated for at least two thousand years. All of them were brought to the New World, if not in the very first ships, then surely within the first dozen years. Most of the plants that the colonists had known in Europe were promptly tried out in the new land and the majority of them were retained.

Tomatoes can hardly be considered as part of the pioneer garden. The tomato is a South American plant and, like the potato, it reached North America by way of Europe. Here it attracted little attention until after the middle of the past century. An old name for the fruit was "love apple." For some reason the plant had a sinister reputation for a long time. I remember that in my youth tomatoes were grown in most gardens and were eaten and enjoyed. There were many old people, however, who insisted that they were a cause of cancer. Tomatoes were fully accepted only in recent years.

Unlike many of the nomadic tribes of the Western Plains, the Indians of the eastern United States were sedentary, occupying the same village sites and agricultural lands for long periods of time. Hence it was that they became rather advanced gardeners and horticulturists. Their crops were few: corn, beans, squash, pumpkins, artichokes, and tobacco. They had, however, many varieties of corn, beans, and squash. All of the foregoing were New World plants, entirely unknown to the white man. But the pioneer was quick to borrow from Indian agriculture. He not only immediately adopted corn as his own but he quickly added the beans and cucurbits of the friendly tribesmen to his garden. From the Iroquois Indians he adopted the field pumpkin and many sorts of squashes and hard-rinded gourds. Some of the hard gourds, after the soft pulp was removed, made excellent vessels, containers, and dippers. The open-minded housewife availed herself of these as part of her household equipment. Indeed, I can remember when the gourd dipper was used in our home.

The bean is another exceedingly important plant which the Indians contributed to our agriculture. Europe had long grown a plant of a related genus called bean, horse bean, broad bean, or Scotch bean. This plant is grown by the Scotsmen of Canada, but has never become one of our American farm crops. With the exception of the Lima bean, all the beans of our gardens and fields, including such diverse forms as the bush beans, the pole or climbing beans, and the string or snap beans, are types of a single species popularly known as the kidney bean. The tribesmen of the Six Nations were particularly good farmers and grew beans almost as widely as corn. Beans were a most vital addition to their menu because, when the meat supply ran short, as oftentimes it did, beans were a high protein food. The white man realized the value of beans and the Indian staple soon held a prominent place in the farm gardens of the pioneer period. Succotash—beans and corn stewed together—was a true Indian word adopted by the settler and it is now a fully accepted term of our everyday English speech.

The discovering of new sources of food was by no means one-sided. In the early days in the New World, the colonists brought seeds of the five best-known European orchard fruits—apple, pear, peach, plum, and cherry. With the exception of some very inferior wild plums and cherries, the Northeast had no edible tree fruits. The Indian was quick to recognize the superiority of these new fruits from across the sea, and he soon adopted them for his own use. When, in the Revolution, Sullivan's expedition advanced into the utterly unexplored region west of the Genesee River, the army halted in its march to uproot extensive fields of maize and chop down bearing orchards of apple and peach. There is, near the New York State Agricultural Experiment Station at Geneva, a little group of four or five incredibly gnarled and ancient apple trees, which reputedly were the remnants of Indian planting done in, what was for that particular region, prehistoric times.

The discussion of the field and garden crops of the homespun age may be summed up thus: the basic crops of the typical New York State farm of today are essentially the same as those grown at the opening of the nineteenth century. A few crops—more specifically flax, peas, and the root crops—are nearly extinct and some others, most notably alfalfa, alsike clover, and some new grasses, have been added. There have been some very marked horticultural improvements, including better varieties of grapes and small fruits, but these changes have hardly been radical enough to be called sweeping or revolutionary. In an agricultural world where almost everything else has been changing with bewildering speed, the crops of northeastern America have remained relatively stable.

If my great-grandfather, who more than a century and a half ago began his labors on the farm where I dwell today, were somehow permitted to return to the scene of his life's work, he would surely be amazed by the mechanization of all farm methods. He would find almost no operation that would be familiar to him. But if he went for a walk across the unchanging fields which he knew so well, he would feel very much at home. The crops are so little different from those of his time. He would see the wheat, corn, oats, barley, timothy, red clover, June grass, and white clover growing as they did in the days when he walked among them. He would note the quack grass, wild mustard, white daisy, yellow dock, and bull thistle—the same weeds that plagued him —and perhaps he would bring home a handful of alfalfa and anxiously inquire concerning this new weed which was threatening to take over the farm.

CHAPTER XI

Living off the Wilderness

WHAT manner of man was the settler who went forward to possess the wilderness? He was not of any particular class or condition, but typically he was young and he had a woman by his side. To both, all the world seemed before them and their hearts beat high with hope. They went a little way—as a rule, only a *very* little way—beyond the most advanced line of settlement and took possession of the land for which they had bargained. Quite contrary to much popular misconception, there was never a time in New York State (or in eastern America, for that matter) when a man might simply march into the wilderness, select a pleasant plot of land that pleased him, and declare, "This is my own by right of discovery and possession and my title is secure." There never was any such thing as literally free or open land, although it is true that on the very forefront of settlement the price per acre was very low. In the early years of the colony, the pioneer must lease his holdings from one of the Hudson River patroons. He could not purchase his leasehold because it was the established policy of the proprietors not to alienate their land by sale.

A generation or two later, he could purchase his farm from one of the patentees, some of whom held vast tracts of land under letters patent from the English crown. Ultimately, nearly all the land between the Hudson River manors and the Fort Stanwix treaty line was granted to certain favored individuals who paid some sort of feudal "peppercorn" rent. The proprietors of these patents divided them into lots. The lots were subdivided into family farms and sold to land-seeking settlers. The sale of these tracts resulted in the foundation of some of the largest fortunes in America.

West of the region that had been distributed as patents was that which came to be known as the Military Tract. This, in very large part, was allotted as a bonus or award for military services in the Revolutionary War. The settler might have been fortunate enough to have earned a warrant entitling him to certain assigned plots of land, or, lacking this, he might purchase a warrant from some holder who thought it wiser (as very often happened) to sell his interest than to try to develop it for himself.

Finally, when the more easterly regions of the state were occupied, the new settlers had to press on west of Geneva and bargain with Phelps and Gorham or their successors, particularly the Holland Land Company. This is a very sketchy outline of New York State land titles. It is given solely to emphasize the fact that there never was any such thing as really free or open land available by right of settlement or possession.

In those great days when men were chopping farms out of the wilderness, there was one question of pressing importance to them. To what degree could they hope to live from the resources provided by the wilderness during the first one or two years. Merely getting a shelter for himself and family was not the pioneer's biggest job. Compared with the task of clearing sufficient acreage for the crops he needed, his building plans were a minor undertaking.

As has been set forth in an earlier chapter, the pioneer entering upon his woodland possessions in the spring and exercising the utmost diligence and skill might, at the most, hope that July of the next year would find a ragged little field of wheat turning golden around the blackened stumps, and in October he might husk his first meager patch of corn. So too, the heroic woman who shared his fortunes might anticipate that there would be the beginnings of a garden with some of the old vegetables brought across the sea and some of the corn, beans, and squash adopted from her Indian predecessors.

So far as game is concerned, I believe there is a vast amount of misapprehension. Fabulous tales, which are little more than inventions, have floated around and become an accepted part of our folklore regarding hunting conditions. If we are to credit these fairy stories, the new settler (if only he were on the extreme front line of occupancy) had only after supper to take down his gun and go a half-mile from his clearing and within the hour he could almost surely return with a fat deer. Or, if it happened that his family were for the moment surfeited with venison, he might offer them a change of menu by bringing back a brace of partridges or a half-dozen gray squirrel, or what rabbits he could conveniently carry. Or perhaps he had a fancy for fried trout for breakfast. He needed but to walk down to the nearby brook and bring back what he wished, strung on a willow twig.

Tales such as this are purely fanciful, and yet they have a place in the thinking of a great many people. As a matter of record, there is nothing in the writings of men who were a part of this period that lends any support to the thesis that there was ever a time when the unspoiled wilderness so teemed with fish and game. There are at least two men who were competent observers, who came into the region of east-central New York with the very first wave of immigration, and who have left in sober prose fairly extended accounts of conditions at that period.

One of them was Judge William Cooper, best remembered as the father of James Fenimore Cooper, the novelist, but nonetheless richly entitled to be esteemed a literary gentleman in his own right. Judge William first came to the Otsego Lake country in 1785 in advance of the very earliest colonists. A year later he was instrumental in settling some 40,000 acres of land which he had acquired in one of the big land deals characteristic of that era. From then until the close of his life he was a resident of Cooperstown. In 1807, an Irish attorney, one William Sampson, Esq., wrote Cooper posing a number of detailed questions relative to the conditions prevailing in America, especially as to the soil, climate, forests, and animals, as well as to the economic opportunities that might offer themselves. In reply, Cooper wrote him what is called a letter. Two years after Cooper's death this "letter" was published under the title *A Guide in the Wilderness*, and, as since republished, it runs to forty-five closely printed pages. As a source book concerning what life was like in east-central New York in the latter years of the eighteenth century, it is unique and priceless. One must recognize the Judge as an honest man of keen observation and sound common sense who was very definitely a part of what he told. To pick flaws in such a contribution may be ungracious meddling, but it is interesting to note that he did exhibit the unscientific credulity of his time by unreservedly subscribing to the fable that the cow might mate with a moose and bring forth a strange offspring with some of the characteristics of both species. This oddity could not be restrained by fences and had the trait of browsing on shrubs and branches like the moose rather than eating grass like the ox. This was a day when the scholarly biologist was not a member of society, and it is easy to pardon the Judge's acceptance of a common apocryphal tale.

Barring two or three slips such as this, Cooper's dissertation seems thoroughly authentic and down to earth. Certainly he says nothing to indicate that the newly-come settler would find it easy to supply meat for his table by hunting. The whole gist of his narrative is to the effect that the early years of the colony were a desperate struggle for existence. The second winter of the Cooperstown colony, the inhabitants endured not only the usual hardships and privations but were reduced to what was hardly less than extreme famine. I have before referred to the miraculous draught of fishes which relieved their most pressing wants. Clayt B. Seagears, of the New York State Conservation Department, assures me that even today at the proper place and moment this fishing exploit might be duplicated on some of the streams flowing into the Finger Lakes. Then he hastens to add that, despite an occasional episode such as this, the pioneer could not expect to keep meat in the pot consistently—a conclusion supported by all credible testimony.

Judge Cooper enumerates only three "noxious" animals of the countryside as he knew it. These were the bear, the wolf, and the panther. He accuses the bear of making away with pigs, and doubtless this was true of lambs and possibly calves as well, but he makes no mention of any injury to humans. In corroboration, I quote from a letter from Seagears, "Despite thousands of stories, I do not believe that there was in New York at any time any animal that was a menace to the settler or his family." This sets forth the belief of a particularly competent gentleman and it is precisely in line with my own thinking. In all conscience, the pioneer had hardships, privations, and special dangers enough but fortunately he and his family were spared any fear of injury from wild beasts.

A second source book which deals with the same place and time as did Cooper and which remarkably verifies his conclusions is the *Reminiscences* of Levi Beardsley. Beardsley was born in 1785, in Rensselaer County, near the Vermont Line, but when he was five years old his family went to the new country in what is now the township of Richfield in Otsego County. Here he grew up, on the extreme advance edge of the frontier, under conditions that seem to have been about as primitive as anything that might be imagined. His early home was a log house with a dirt floor and a bark roof which leaked so badly that when it rained the scant family belongings must be moved from one place to another.

Ultimately Beardsley became a man of importance. He was for eight years a state senator for his district, and for some years president of that legislative body. According to the standards of his time, he was a far traveler and he had a first-hand acquaintance with many of the political figures of his day, including Henry Clay, Martin Van Buren, and Lafayette. In 1852, a date still within the homespun age, he wrote his *Reminiscences,* and the opening chapters dealing with his Otsego County boyhood have the ring of simple authenticity. His locale and Cooper's were almost identical and the two narratives corroborate each other with the difference that Cooper, the grand seigneur, wrote of what he saw and heard, while Beardsley, child of a wandering frontiersman, wrote of what he himself experienced. Neither of them indicates that the fish and game of the new country was any real assurance against hunger.

What were the animals that were considered hopefully as contributions to the family larder? In popular thinking at least, it was venison that figured most prominently in the woodsman's food supply. The common white tailed or Virginia deer were native everywhere, but it is not certain that they were numerous. The half-abandoned marginal hill lands of our state offer a far better range today than they did when in unbroken forests, and there are extensive regions that probably have a larger deer population now than they carried in the days of the Iroquois.

In those times, however, the deer hunter was entirely unhampered by game laws. Deer were commonly hunted with hounds. Beardsley writes of this practice in much detail. Also, at night deer might be decoyed within range of the rifle by burning flares of white birch bark. While many settlers ate venison now and then, it was probably not possible to get it with

Deer Hunt

enough regularity to make it a major source of meat. There have come down, too, many tales of near famine on the far-advanced lines of settlement.

Competent naturalists say that the elk was not found in New York unless it might be as an occasional visitor in some of the southwestern counties. Cooper mentions the "elk-deer," but this again may have been a slip on his part. Bears were occasionally shot and their skins made a superb bed covering on winter nights. Their meat was freely eaten when available, and Cooper writes that "their hams dried and smoked are preferable to smoked beef." Rabbit, woodchuck, raccoon, muskrat, beaver, and probably the lowly skunk appeared on the menu whenever the proprietor was fortunate enough to get hold of them. To quote Cooper again, "The squirrel and the hare follow the plantations and are seldom seen far from improved land." Rather surprisingly, he adds: "Rabbits are not very plentiful."

There is not even the tradition of the occurrence of buffalo in New York. The wood buffalo, a subspecies of the plains bison, once roamed Kentucky in great numbers. It was common in Indiana and competent zoologists say that its remains have been found in both Ohio and Pennsylvania, but there is no evidence that New York State was included in its range. Seagears suggests that possibly the buffalo appeared now and then in Chautauqua County, but it is safe to say that none were ever killed there. Such a mighty feat would have created a legend that would never have been allowed to perish. We have deer stories and bear stories and wolf stories aplenty, but no legends of the buffalo. In any case, the question is academic because, unlike the deer, the bison is a timid brute that withdraws before the presence of the permanent settler. Donald Culross Peattie says that by 1810 the buffalo had been pushed west of the Mississippi. A magnificent beast, whose countless herds in their heyday have been estimated to number seventy-five million, would, save for the devotion of a few farsighted conservationists, today be as extinct as the dodo bird.

Cooper enumerates in considerable detail the game birds of his locale. The wild duck he calls "common," but he lists the wild goose as unusual. He characterizes the partridge as "abundant." Quail were rare and followed rather than preceded the clearing of the land. Cooper also names the woodcock, snipe, and plover, and of course that incredible miracle, the passenger pigeon. Probably Cooper's was not so much an effort to make an inventory of the food resources of the new country as it was an attempt to tell an Irish gentleman what sort of sport he might expect if by chance he should go ahunting in the American wilderness.

The most astonishing, and economically the most important, of our birds was the fabulous passenger pigeon. Trained ornithologists like Audubon and Alexander Wilson have left records of their unbelievable numbers. Any rational calculations such as Audubon attempted places their number in billions rather than any lesser enumeration. Always men marveled at their spring migrations, which darkened the sun and cast shadows for hours at a time. Strange tales have come down to us concerning them. Sometimes their flight was so low that a man standing on the brow of a high hill and swinging a pole at random knocked them down by scores. One story relates how a hunter discharged a shotgun into the rushing torrent of birds and picked up twenty-six pigeons. The farmer who chanced to be on one of their flightways in April might take as many as he wished and, even better, if there was a nesting ground in his vicinity he might gorge himself on the succulent squabs. Their disappearance within a very brief span of years remains the great unsolved riddle of ornithology. To my father they were a perfectly familiar phenomenon but I have never seen a single specimen!

If the fish and game resources of the wilderness provided only a precarious basis for the family dinner table, the vegetable kingdom promised even less. The list of wild plants that may be eaten is a long one, but I cannot feel that those native to the New York State forests were fundamentally important. There is a class of plants that the housewife designates by the generic word "greens," and which the economic botanist refers to by the descriptive term "pot herbs." Of these the best known and most commonly used is the dandelion. This plant, while now naturalized everywhere, is of European origin and was not available on the front line of settlement. Purslane, or colloquially "pussley" of old gardens, is also (as very many other plants are) a European migrant. The common nettle, cowslip, or marsh marigold, and milkweed are American species that have gained a good repute as pot herbs. My own mother, within my memory but before every country store had fresh vegetables all winter, used to speak of the "six-weeks' want," meaning the early spring period when the old vegetables and apples were gone and nothing new was available. No wonder the housewife scoured the spring fields for something green to cook. Pot herbs were a change and probably they corrected certain vitamin deficiencies, but after all they were mainly water and their caloric values low. Again, these familiar greens were commonly plants of old fields and gardens and probably would not be available to the pioneer homemaker.

There are two wild plants that deserve special mention. As has been said before, Judge Cooper writes how in the year of the great hunger the Otsego Lake colonists resorted to the bulbs of wild onion or leek. This is by no means the only instance when this malodorous food was used. The leek grows in dense clumps fully occupying the ground, and a few square yards would furnish a considerable weight of material that might seem good to a truly hungry man. In Judge Brown's narrative of the Schoharie settlement, the very nutritious groundnut seems to have been a providential addition to the food supply at a critical time. Brown's story gains an added trustworthiness when we read Asa Gray's statement as to the usual habitat of the groundnut: "rich alluvial valleys and river banks." This precisely describes the particular type of land on which the fortunate Palatines of Schoharie lighted. It seems certain that the food possibilities of this native plant were widely recognized.

Then, there were certain native fruits, very few of which would compare at all favorably with our cultivated sorts. There was the red raspberry, black raspberry or blackcap, "long" blackberry, elderberry, and, of course, the wild strawberry, of such diminutive size but superlative flavor. In some localities there were two or three species loosely grouped together as huckleberries (New York State usage) or blueberries (New England speech). Again, all of these are fruits of newly cleared lands and abandoned pastures, and the family on the farthest fringe of settlement might find them almost nonexistent. The common wild grape of New York is the river-bank grape whose clusters are small, sour, and seedy, so that only with plenty of sugar (which the pioneer wife surely did not have) would they make a palatable jelly. The fox grape is common in New England but rare in New York. It is an exceedingly variable species and some vines bear fruit that is very edible. So far as tree fruits are concerned, there may be mentioned wild plums and three sorts of cherries: pin, black, and choke. These may all be eaten, but they are small, difficult to gather, and hardly deserve to be reckoned among the food resources of the untouched forests.

Of course, there were nuts: hickory, butternuts, and in a few favored localities, black walnuts. In parts of the state there was the eagerly-sought-for chestnut. Here and there were the bushes that bear hazelnuts and almost everywhere was the tiny beech-nut, almost too small to be worth the time of a man but a choice food for wandering turkey and squirrel. All nuts are a delicious and concentrated food, but it must be said that a good crop of nuts occurs hardly one year in three and they would be most un-certain as even a minor addition to the menu. From boyhood experience I can testify that the acorn of the white oak, the "sweet" acorn as we called it, can be eaten, but to do this re-quires the appetite and digestion of a growing boy. Much more important, in regions where a large part of the forest growth is oak, the fallen acorns, collectively known as "mast," furnished excellent feed for hogs, allowed to range the woodland.

The question as to how far the newly arrived pioneer might depend upon the food resources of the wilderness adds up to this: so far as consistently keeping meat in the pot, the prospect was highly uncertain. If he was a skilled hunter he might now and then kill a deer or a bear and live well for a time. Once in a blue moon there might be a miraculous draught of fishes, such as so greatly helped the Otsego pioneers at a time of crisis. Perhaps, by happy chance, there might be a great flight of pigeons or they might even establish a nesting ground at some convenient distance so that for the moment the settler might have food and to spare. But these fortuitous happenings could not assure him against want during the bleak months.

So far as living off the vegetable bounty of the wild country, the prospect was even worse. Remember that from the day he cut down his first tree, it would be at the least a full year before the pioneer could hope for even the smallest return from his agricultural operations, two or three years before he could reap any considerable acreage. When it comes to providing a living for a family, a corn patch of a couple of acres and perhaps a little field of wheat among the stumps is worth more than all the roots, herbs, and fruit that can be gathered from the whole virgin countryside. So too, a milch cow, along with a sow and a litter of pigs, is a better promise of good eating than all the beasts of the forest that the settler might hope to bring to earth with his trusty flintlock rifle. If, when he made his advance into the new country, the pioneer lead with him a cow, she would certainly be able to browse her living for the first summer, and perhaps he could get enough forage to keep her alive over winter. If he was so opulent that his impedimenta included also a sow, she could run wild and bring up her brood as razorbacks.

The Pig in the Pioneer Yard

It is my firm conviction, however, that for the first one or two hard years the pioneer on the front line could survive only by purchasing some of his food from earlier settlers who already had a surplus of bread stuffs for which they sought a market. In saying this, I know that I am denying the most dearly cherished traditions of the romanticists, but, so help me, I cannot do otherwise.

CHAPTER XII

The Wooden Age

THERE is an old classification device whereby archeologists and historians divide the long span of the years behind us into ages, each with its distinctive name indicative of the cultural and economic development of that particular period. The earliest of these, the Stone Age, was the childhood of mankind. Some centuries later men, still growing wise as the generations passed, mastered the difficult art of making iron from ore. So came the Iron Age, in the midst of which we seem to be and which as yet shows no sign of passing. Along with the Iron Age, we have also the Machine Age, which is particularly typified by the assembly line and by wonderful automatic devices. And now we are assured that just over the horizon waits the Atomic Age, which holds no man knows what, unless it be the danger that civilization itself be "hoist by its own petard."

I do not know that it is written in the books but I would like to christen and offer for official recognition yet another age: the Wooden Age, which flourished particularly in eastern North America and which attained its finest flowering as a development of the later years of the homespun era. There were two very good reasons why this age waited until rather late in history and why it reached its richest fruition in eastern America. One was that fine craftsmanship in wood was impossible until we had tools of many types made of steel which could be sharpened to a razor edge. The second reason was that by the time woodworking tools became available, the forest resources of the Old World had largely been depleted so that men must build homes of stone or brick and roof them with thatch, slate, or tile. However, long before the pioneer reached America, he had learned to forge keen tools of steel. In the land to which he came he found wood of many sorts suited to his purpose and in what seemed inexhaustible abundance. It is a tribute to his resourcefulness that he so promptly turned to wood as a building material.

It is certain that at the very beginning, before the discovery of a better way, the pioneer made some effort to copy the architecture of the homeland from whence he came, but very soon he was rolling up logs to form the structure that was to become the universal home of the pioneer along the whole Atlantic seaboard, the log house. Usually within a generation a frame dwelling would replace the log house. Most of these were simple farm homes, but here and there, as for example the shipmaster's dwelling of the New England coast or the manor houses of the Hudson Valley, were built homes of wood which, after a century or more, still stand, dignified, lovely, and serene. All in all, we have a good many worthy houses built by old time boss carpenters whose most indispensable tools were the chalk line and plumb bob, along with the broadax, the framing chisel, and the long jointer plane.

The zenith of the wooden age, which fell rather precisely about the middle third of the nineteenth century, was represented by the great traditions of New England shipbuilding. Competent artists have declared that a full rigged ship under sail is the most beautiful thing that the imagination and hand of man has ever created. Somehow or other, the Yankee seems to have been gifted with a strange genius for building and sailing ships. He built them out of native oak and pine with tough elm for the keel, and gave them fanciful or whimsical names like North Wind or Water Witch, or such proudly glamorous names as Flying Cloud or Sovereign of the Seas. They were the swiftest craft that ever fled before the wind, and Yankee captains made the American flag familiar in every port of the world.

Portland, Newburyport, Portsmouth, Salem, and many other less widely known towns are noteworthy for the stately homes built by the great shipmasters in the halcyon days of wooden sailing ships. After all the years they still show in every line excellency of workmanship. Most typically they were topped by a glazed cupola and outlook known usually as "the captain's walk," but sometimes by the more sinister term "the widow's walk." The secret of the beauty and permanency of these houses is that the men who built them were able to command the services of the same men who built the ships, and the ship carpenters were the most expert craftsmen of their time. They had to be such if they were to build wooden vessels that could sail to the Seven Seas and come safe home again.

But, after all, the big houses of the shipmasters and the manorial homes of the wooden age were relatively few and far between. The spirit of the wooden age is best expressed as it relates to the common man—the frontiersmen who were creating a substantial civilization where there had been only wilderness. Any enumeration of the pioneer handicrafts grows into a long list and some of them were highly specialized. As the army of occupation pressed slowly westward, it was the man with his ax and his ox team who represented the most advanced skirmish line, but close on his rear marched a supporting corps of craftsmen of diverse sorts. It is true that the omnipresent blacksmith worked in metal and it may well be true that he, above all workmen, was indispensable in the conquest of the frontier, but with this single exception the mechanical activities of the community were predominantly crafts in wood.

Of all the occupations of the homespun age, priority in numbers at least goes to the carpenters. In that oft referred-to Cattaraugus County of 1855 there were 267 men who so classified themselves. In number they made up approximately one-third of all the men who claimed a specific craft. It is not difficult to explain this array of carpenters in a single, almost entirely rural, county. The countryside was not yet mature. During the previous decade the county had gained some ten thousand inhabitants. New families were still establishing homes and must be provided with houses. The original log houses were being replaced by frame dwellings. All not yet fully settled regions were experiencing a perpetual building boom. Moreover to build a house in those days required several times as many man-hours as now. The boss carpenter of a century ago had never heard of the balloon frame. All his work was characterized by its massiveness. The posts, main beams, and cross timbers were shaped out of the log with the broadax and adz. The siding was planed by hand—a job commonly turned over to the young fellows learning the trade. The window frames, sashes, and doors were built by hand on the job. The floors were of broad boards, planed smooth, then tongue-and-grooved by hand. In a word, every house and barn was a hand-tailored job. No wonder that to build a plain but reasonably commodious farm home was to employ a gang of five or six carpenters all summer.

If the house and barn carpenters were numerically and economically the most important craftsmen in wood, it would seem that second place must go to the coopers. The two crafts are really a good ways apart. To compare the ordinary run-of-the-mill carpenter with a "tight" cooper is like comparing a lumberjack with a skilled cabinetmaker. Both work in wood, but here the resemblance ends. To go into the woods, cut down a big white oak tree, and, aided only by a few simple tools, convert it into barrels which would hold liquids or into butter tubs which would meet the test of being "brine tight," called for a high order of skill. Granted that both the smith and the carpenter were indispensable, it is hard to imagine how the pioneer could have carried on without the cooper. The cooper made an amazing variety of necessities. He made flour barrels, cider barrels, and pork barrels. He made the meat tubs, the wash tubs, and the butter tubs. He made the sap buckets and the old oaken bucket that hung in the well. He made the pails the farmer held between his knees when he milked his cows, the pails from which the calves were fed, and the kitchen pails. He made the oaken dash churn found on every well-ordered farm. In a word, he made everything formed of staves drawn together with wooden hoops.

There were two examples of the cooper's art that deserve special mention. One was the sap bucket and the other the white oak butter firkin. There was a time when, in New York and

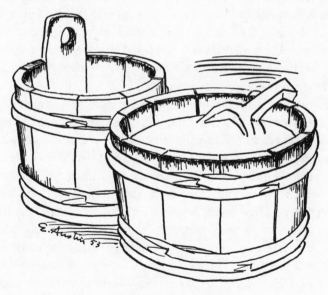

Sap Bucket and Firkin

New England, not tens of thousands but literally millions of wooden sap buckets were required, every one made by hand in a cooper shop. They were indispensable in the making of maple sugar, which held so large a place in the economy of the pioneer era. They were shaped, not for finish nor beauty, but solely for utility, and not one unnecessary stroke went into their construction. Almost invariably the material was the very best old, free-splitting white pine, although cedar was sometimes used. Neither the inside nor outside of the staves was touched with any tool, but left with the grain showing as rived from the block. The edges of the staves, however, must be beveled and jointed with almost perfect accuracy, and the bottom must fit the chine, the groove cut to receive it, with the same preciseness. If there was any object in the world that entirely failed its purpose, it was a leaky sap bucket. Each bucket had commonly six light wooden hoops. About the middle of the nineteenth century—a period when the cooper's art was still in flower—sap buckets were sold for what seems the incredibly low price of six cents each. Such a figure was possible because the splendid raw material at that date cost very little, the whole rural economy was based on a low wage and price level, and the coopers who were building sap buckets were skilled artisans who made no false motions.

Before the onslaught of the machine age, the rural cooper beat a slow retreat. His last stand was the making of oaken butter tubs and butter firkins, an industry still active at a date when, otherwise, coopering had been reduced to assembling machine-made staves and headings. Within my own immediate observation, at least two coopers plied their trade until the early years of this century. They were, however, specialists, restricting their work to one narrow field: the production of oak butter tubs which they built from standing tree to finished firkin according to traditional methods. The raw material was large, free-splitting white oak logs which were sawed into blocks thirty-three inches long. These blocks were then rived in staves of approximately the correct thickness. At this stage of preparation they were known as bolts. The bolts must then be seasoned for several months before being shaped and dressed with precision accuracy to make a firkin which emphatically must be "brine tight." The hoops were shaved from hoop poles, young saplings of hickory, white ash or white oak, preferably from one and a half to two inches in diameter. Such a sapling could be split and shaved so as to make four, and sometimes six, hoops. I have been assured that to cut the notches—the means by which the ends of the hoops were locked together—was the neatest trick the apprentice cooper had to master.

There was this essential difference between the making of sap buckets and the butter tub: the sap bucket was a mass production job which must be done cheaply. Sap buckets were turned out in great numbers and must be produced at the lowest possible price. The butter tub, on the other hand, was designed for the marketing of a valuable product where the external appearance of the package was important. The butter-firkin cooper put into his work pride of craftsmanship. The bucketmaker was quite content if only his product proved watertight. He dressed the staves only on the edges. But the butter-tub artist planed his staves not only on the edges but on both sides as well. Before

The Cooper

the firkin was complete, it was made very smooth and given an outside coat of varnish. The handmade oak butter tub had trademark value in the markets of fifty years ago.

When, in the 1840's, the first shipment of milk was made into New York City from Orange County over the Erie Railroad it was carried, not in forty-quart tin cans, but in oak churns, the most practical containers that could be assembled for the purpose. In the great days of the Syracuse salt industry, around the middle years of the past century, the making of salt barrels was second only to the business of boiling salt.

In the years when the cooper's craft was in flower, it was recognized that there were two degrees of skill within the guild. There was the "loose" or "slack" cooper, a less expert workman whose ability sufficed only to make such vessels as need not be watertight. His field was flour, salt, or fruit barrels, measures or containers. In a day when there were no paper sacks and no burlap bags, barrels were used for shipping various commodities. Judge Cooper writes how at an early period his Cooperstown settlement sent to market important amounts of maple sugar in hogsheads (a commonly used term for a greatly oversized barrel). Doubtless the hogsheads were supplied by local coopers, but it was not at all necessary that they be watertight.

The other class of coopers—aristocrats of the art—were the "tight" coopers, meaning men whose skill of hand and eye was such that they could shape, dress, and bevel the staves so accurately that the resultant vessel would be as tight as a new tin pail. These were the men who built the cider barrels, the whiskey barrels, and the meat casks.

I have heard of yet a third classification, "white" coopers, but I do not know the particular connotation of the term. Reputedly this was the name applied to workmen who specialized in containers not of the cask or barrel shape, but with straight, tapered staves to make what we call a tub. Dash churns were sometimes of this type, with the larger diameter at the bottom. I do not know how much of this is fable and how much fact. The men who might have spoken authoritatively are all dead and they have left behind them only a vague, half-remembered tradition.

Hook Splicer

CHAPTER XIII

Joiners and Cabinetmakers

THE term "joiner," in the craftsmanship sense at least, has become uncommon in our vocabulary. In the era that concerns us here, the word was in common use.

At the date of that first occupational census, 1855, the term retained an everyday significance. It is interesting that in Cattaraugus County some 257 men called themselves carpenters, while there were no less than fifty-nine others who announced themselves as joiners. I doubt if there was ever any clear-cut distinction between the two crafts. I believe the joiner was a glorified carpenter who held himself a cut above the rest of his craft, just as the tight cooper deemed he belonged to a different order from the slack cooper. The man who hewed the posts and beams with ax and broadax, mortised the timbers, nailed on the siding, shingled the roof, and, in general, did the rougher work on a house or barn, was content to be regarded as a carpenter. The man who built the interior finishings, the paneled doors, window sashes, stairs, and railing, and who molded the wainscoting and made the fanlights, these men preferred the status of joiner. Probably the term is almost never used nowadays because the craft itself is practically extinct. The work, once done slowly and patiently with keen tools and skill of hand and eye, has universally been taken over by automatic machines.

In the year when Cattaraugus County had 257 carpenters and fifty-nine joiners, there were also forty-four cabinetmakers. If the line between carpenters and joiners was uncertain, it is surely hard to say where joinery left off and cabinetmaking began. In common spech and in lexicon definition, the cabinetmaker was primarily a builder of furniture, but just what constitutes furniture? By general agreement, tables, chairs, bedsteads, chests of drawers, trundle beds, cradles, and movable household equipment were furniture.

As soon as men had achieved anything like a secure civilization they began to lavish care and craftsmanship upon the utensils of the house. For centuries Europe has had furniture inlaid and carved with most intricate skill. Hepplewhite, Sheraton, and Chippendale have a secure place in history. Some early masterpieces were brought to America and were reproduced by copyists. By the time of the Revolution these fine pieces had a place in the pretentious homes of our seaboard cities, in the plantation homes of the South, and the manor houses of the Hudson Valley. They symbolized wealth, and perhaps culture, but hardly comfort. I wonder if anybody ever ventured to loll at ease in a Chippendale chair. My interests lie with men who wrought at their tasks so that the rapidly multiplying pioneer homes might be provided with the fundamental requisites for comfortable and orderly living.

For tables, cupboards, and chests the early cabinetmakers worked most frequently in pine, because it was light in weight, durable, and did not warp. It was readily available in wide boards, and, most important of all perhaps, it was soft, fine-grained, and peculiarly satisfactory for shaping with tools. Another wood greatly esteemed by the pioneer craftsman was cherry. Wild cherry was a common and sometimes large tree of the Northeast. It was harder than pine, but fine-grained. It took a beautiful finish and was not too difficult to work. But probably the greatest reason for the popularity of cherry was its natural reddish-brown color which, to some degree, approximated the precious mahogany and rosewood so utterly beyond the attainment of the typical cabinetmaker on the American frontier.

White pine and cherry were fine for tables and closets and cupboards, or any use where strength was not a prime requisite, but when it came to chairs and bedsteads, the cabinetmaker turned to hard woods, especially oak, ash, and maple. Probably the choicest wood to which the frontier workman had access was bird's-eye maple, a rare variety of the hard or sugar maple. Trees carrying this variation were few, and those differed greatly in the extent to which the curlicue graining was developed. This beautifully figured wood appears frequently in the best of old-time joinery.

I deem it one of the greatest blessings of my life that as a college boy I sat for a year, five mornings each week, under the lectures of I. P. Roberts, Cornell's great teacher of agriculture. When in reminiscent mood he would unfold a tale which, better than any other I know, symbolizes the rigors of life on the advanced frontier. It was 1812 when I. P. Roberts' Grandfather Burroughs, with the family possessions in an ox-drawn wagon, made the trek west to the forests of Seneca County. When his grandfather came to what he was told would be the last sawmill on the way, he halted and bargained for a single, very wide, pine board. This purchase, supported on wooden pins driven into auger holes bored in the logs that made the house, constituted the Burroughs dining table for a considerable span of years. Surely, when the clan surrounded it at mealtime it might be said with literal truth that they were gathered about the family board.

The history of the grain cradle, together with a description of the various types and their use is most properly a part of the story of how the pioneer brought the harvest home which I shall tell later. In passing, however, a word ought to be said in behalf of the men whose skilled craftsmanship made possible this indispensable implement of the homespun age. The occupational census of 1855 does not specifically enumerate the cradlemakers but lumps them under the general title of Agricultural Implement Makers. At a rough guess, however, there must have been in the nineteenth mid-century at least a quarter of a million cradles on New York farms and at least a million in the northeastern states, and the construction and maintenance of these must have employed a considerable number of highly skilled artificers in wood. I remember in Otsego County a little milldam and a shop that was locally spoken of as the cradle factory. It may be that three or four men worked here and probably their activities were not confined to cradles. It was an industry, however, where there was small advantage in any attempt at mass production.

At Mineral Springs, in Schoharie County, New York, lived a cradlemaker, one Christian Bouck, who wrought at his craft until within the easy memory of men still living. Doubtless few of his productions ever went more than a dozen miles from the little shop where they were made, but within the bounds of his bailiwick his work had an enviable reputation. A good cradle is an object of wonderful construction. It must be strong enough to stand the strain and tug of laying heavy wheat and yet it must be light in weight or it is a "man-killer." It must have that symmetrical balance we call "hang." The old-time, expert cradler was as finicky concerning his cradle as his modern descendant regarding his golf clubs. Strength and lightness and "hang" could be achieved only by a combination of the best ash and hickory with skillful workmanship. Cradle "fingers" must, of course, have a certain degree of curvature, and Bouck would go where big ash trees had been cut down, dig around the stumps, and from the spreading roots split fingers with the desired natural crooks. From his bench came a finished cradle, in itself an artistic triumph comparable with a Stradivarius violin. Men hereabouts in my native Schoharie used to declare with pride, "That's a genuine Bouck cradle."

Wagon and sleighmaking may properly be bracketed together because their methods and materials were so similar they were in reality one craft. Let it be noted in passing that in some instances the earliest pioneers had no draft animals. This seems to have been true of the Plymouth colony, and almost a century later when the Palatines made a mass migration of six hundred souls into the Schoharie Valley they had neither horses nor oxen. They were particularly fortunate in that their lands included several old fields cleared by the Indians, and these were prepared for planting and sowing simply by the use of broad, heavy hoes and mattocks. Such a procedure represented an amount of backbreaking toil which to us today seems incredible, but for two or three years they grew their bread grain (corn and wheat) that way. Just as soon, however, as the pioneer had achieved a measure of security he had oxen and, to a less extent, horses, and then he needed wheeled vehicles.

We would like to know something regarding the craft whereby the plain farm folk of the wooden age were provided with carts and wagons for the many uses on the farm or for trips to "mill or meeting." That old phrase, "to mill or meeting," has a definite quasi-legal connotation. In the days of toll roads it was commonly agreed that the payment of toll must be waived if the traveler could show that he was on his way to either gristmill or church.

It would seem that the wheelwright of the pioneer era put a major part of his effort into two-wheeled carts rather than the conventional four-wheeled wagon. There are at least two good reasons for this: one was the greater simplicity of cart construction. There is quite a bit of connecting gear, hounds "crotch," reaches, and kingbolt, between the front and rear axle of the four-wheeled wagon. Possibly the dominant reason, however, for the prevalence of carts is that the fields where the harvest must be gathered were still studded with stumps and sprinkled with boulders. A cart was easier to get around such obstructions than a four-wheeled vehicle. There was only one pair of wheels to watch. The heavy, two-wheeled oxcart was once almost universal. It lingered to some extent as long as oxen were found upon our farms. While the farmer going abroad alone on matters of business or pleasure most commonly rode on horseback, he sometimes drove a single horse before a light cart. Even within my memory, an occasional man might be seen making his way to town in a single-seated, two-wheeled vehicle, with us called a road cart, but in some regions a gig.

The diameter of the wheels of the early vehicles was notably greater than on similar wagons nowadays. Big wheels have high clearance which enables the driver to straddle a low stump or boulder. A high wheel rolled better over muddy or rough roads. Even in the heyday of the wooden age, the wheelwright had to call on the blacksmith to provide the iron tire without which no wooden wheel could carry a real load. In addition to the tire, a drawbolt and a kingbolt were indispensable in the primitive wagons. Also, the hub had a hole bored in the wood to receive the end of the axle which had been shaped to fit it. In later work, this hole was provided with a cast-iron lining or box. These old wooden axles were lubricated with various homemade concoctions, a mixture of tallow and pine tar being a common formula. In the wooden age axles were universally fitted with linchpins to keep the wheel in place. I believe it was John Fox who somewhere made the dramatic epigram that the breaking of a linchpin in the Cumberland Gap changed the course of American history. It is an arresting statement, but as a matter of fact the loss of a linchpin was about the least misfortune that could have overtaken the westering emigrant. Any woodsman would have corrected the disaster in a matter of minutes. In passing, it may be said that when the ancient linchpin was replaced by a machine nut screwed on the end of an iron axle it was, in effect, an announcement that the craft of the wooden age wheelwright had come to an end.

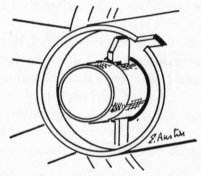

Linchpin

Wagon and sleighmaking demand suitable woods, and, in the years we are considering, the best of timber was almost everywhere available. The workbench builder of wagons (as did his machine age successor) used mainly white oak and white ash, and to a less extent hickory. By common consent the hubs were made of selected tough white elm for the very good reason that it is the hardest to split of all our native timber. Perhaps in our language there is no better example of quietly whimsical humor than Oliver Wendell Holmes's "one-hoss shay" and its eventual tragic demise. Somewhere along the line of his experience the genial doctor must have had real contact with farm woodcraft, otherwise he could never have hit it off so happily and with such truthfulness:

> The hubs of logs from the "Settler's ellum,"—
> Last of its timber,—they couldn't sell 'em,
> Never an axe had seen their chips,
> And the wedges flew from between their lips.

Hubs were made either from small trees or else large limbs of approximately the right diameter. These were sawed into short blocks of the desired length and shaped with the hand ax and drawing knife. Then they must be bored to fit the axle and mortised to receive the spokes. The standard number of spokes was twelve for the front wheels and fourteen for the rear. To space accurately and then work out these mortises in a tough elm hub was a job calling for skill and plenty of patience.

The spokes of ash or hickory were split out of straight grained blocks of the proper length and shaped to desired size and form, using a shaving horse and a drawshave. A spoke auger was used to shape the outer end so that it would fit the hole bored in the rim or felly. This felly was sawed out of oak plank with the very narrow compass saw, which could make a cut following a curved line. The pattern was outlined or scribed, so that the pieces, put together with pin and dowel, would form a complete and true circle. The inner end of the spokes were tenoned so as to fit accurately the mortises in the hub. All in all, there was an appalling amount of labor before the finished wheel was ready for the indispensable tire. Even at the peak of the wooden age, a wagon must have an iron tire as well as a few other small bits if iron equipment. Supplying these, as well as the shoes and braces which were part of a sleigh, was known as "ironing off." At this point, the village blacksmith might take over or sometimes the worker combined the two dignities of artisan in both wood and iron.

It would seem that the handicraft wagon builder felt the impact of the machine age earlier than some other occupations. In such details as turning hubs and spokes and sawing out part of the connecting gear, mechanical power and powered tools offered an almost immeasurable advantage. Some of the old-time builders sensed this and made an effort to adapt themselves to the new day. In this immediate bailiwick and right out in the open country, one Philip Richtmyer maintained a wagon shop for many years. He passed on about seventy years ago, and I do not know that I ever saw him, but some years later I looked over his shop, which was still much as it had been when he laid down his tools. He had a primitive arrangement whereby a single horse, walking around in a circle at the end of a sweep, furnished power for turning hubs and spokes and such other purposes as he could devise. I doubt not that in perfecting this contraption he felt he had made a great forward step, but it was not enough to save him and his occupation from the catastrophic changes that overwhelmed our rural handicrafts during the years immediately following the Civil War.

I have already noted the fact that when power machinery began to displace the old-time cooper, some men at least managed to put off the evil day by becoming mere assemblers of machine-made staves, headings, and hoops. Kurt DuBois told me that this was about the only coopering he ever did after 1875. In precisely the same fashion, some old-time wheelwrights ceased to use native timber and hand methods. Instead, they managed to survive for as much as a generation by becoming assemblers of factory-made hubs, spokes, and fellies, with perhaps some of the connecting gear made on their own workbenches. The resultant wagons were still referred to as handmade, and there was a common, but probably mistaken, belief that they were superior to those turned out by mass production methods. In Schoharie County, Henry (Hankey) Bellinger had a one-man shop behind his house where he made or, more correctly, put together wagons, down until the opening years of this century. When he at length ceased operations, it had been a good many years since he had shaved a spoke or mortised a hub. Thus he made his little business last out his time, but he left no successor. Those few farm wagons that are still made have iron wheels and rubber tires and, with the sole exception of the wooden tongue or pole, there is no single bit of timber in their make-up. Such a change represents an almost incredible evolution which has been in the making for more than one full century. It hardly seems necessary to make place for any discussion of the altogether essential sleigh-making because it was only a subdivision of the wheelwright's art, and without exception the pioneer builder made both wagons and sleighs, so one craft may be considered as embracing both activities.

Our earliest statistics as to the number of men who made a livelihood in the industry leave us with a sense of confusion. In Cattaraugus County in 1855 there were only fourteen men who announced they were wheelwrights. I feel sure that this must have been only a fraction of the true number. The fact is that the necessity of ironing off a wagon or sleigh after the woodwork was completed caused wagonmaking and blacksmithing to be joint enterprises. Within my memory, the blacksmith who could not take on also a job of wagon repairs would have been deemed one who had only half-learned his trade. The wheelmaker's art was a vital part of the fabulous age we are remembering. It prospered in a multitude of little shops scattered over the whole Northeast countryside. Yet it is now wholly departed, leaving only a vague and fading tradition.

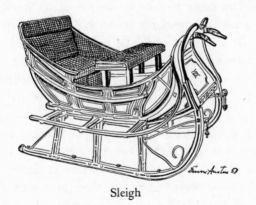

Sleigh

CHAPTER XIV

The Old Millstream

ANY list of the various crafts in wood can hardly omit consideration of "pump logs." Time was when they had an important place in the farm economy. Water—pure, convenient, and abundant—is an essential of living. Often the settler who had purchased and come to take possession of his hundred or two hundred acres of forestland found his family water supply a problem. Of course it could be partially solved by digging a well, anywhere from ten to thirty feet deep, and equipping it with ancient well sweep or the slightly more advanced curb, with windlass and rope. In all the early years the best that could be hoped for was the old oaken bucket that hung in the well. Digging and stoning up a well, however, was no small chore, to be casually undertaken in the intervals between other jobs. Rather, it was a major enterprise. Very frequently, perhaps typically, the pioneer depended upon a nearby spring or running stream. Fortunately, when the countryside was still virgin, water from any convenient brook was no health hazard.

Sometimes it was the presence of a spring which determined

the location of the home. Within the walls of the Witter Museum on the State Fair Grounds at Syracuse a log house has been re-erected with meticulous fidelity to its original condition. The house formerly stood in a remote Catskill Valley, and at the time it was taken down for reassembling I became familiar with its locale. Only a rod or two from the door an abundant spring welled up and slipped away down a little gully. Everything else about the place might be wrong, but there was at least a peren-nial source of water beyond criticism so far as contamination was concerned. I have never doubted this spring was the reason why the first owner pitched his home on the precise spot.

I do not suppose that the housewife of the authentic pioneer era, even in her wildest imaginings, ever dreamed of hot and cold water in the kitchen sink at the touch of a faucet. Nor could she conceive a modern bathroom with tub and shower. I am sure, however, that there were a great many farm women who had vision enough to know that running water beside the kitchen door would be a marvelous help in household tasks. I am equally sure that this laudable ambition would be aided and abetted by hus-bands who in winter time must drive livestock a considerable distance to drink from creek or spring or who must draw water from a well when slopping the hogs.

So it was that far back in the pioneer era there were men working on the problem of supplying the farmstead with run-ning water. A good many farms had springs, but the trouble was their distance from house and barn. The crucial problem was some sort of pipe to lead the water. Our familiar and almost universal iron pipe of small diameter was not on the market in those years. Lead pipe has been known since the days of the Roman baths, but such pipe cost money and the pioneer had little of it. Even if he had been able to purchase lead pipe, the difficul-ties of transportation from the seaboard cities would have been in-superable. With the resourcefulness that seems to have been a part of the age, he hit upon the happy plan of taking small logs and boring a hole through them from end to end; they were then joined to make a continuous pipe.

The raw material for such pump logs was always available in such abundance that it was almost without price. When it was to be had, the favorite wood was white pine because it was soft and easy to bore, and it was commonly obtainable in small, smooth logs of uniform diameter. The logs were joined one to the other by shaping one end of a log into a cone which would accurately fit into a cone cup in the next log. The hole through the log was commonly one and one-eighth inches in diameter and the logs were usually from six to ten feet in length. Pump logs were not to be recommended for long distances and surely not for heavy pressures. There was always the danger of leakage where they were joined; nonetheless this primitive device of the wooden age has the honorable distinction of being the forerunner of all farm water supply systems.

It may be interesting to ask how large a place these wooden water pipes held in our rural economy. Of course, it is evident that there can be no possible statistics. Any figures are at best empirical. The following guess is offered merely as an indication that long before the days of modern plumbing the farmer had buried an astonishing mileage of wooden water conductors. In that oft-quoted year 1855, which must have been about the zenith and also near the end of the pump log era, there were in New York State some 232,000 farmsteads. If, purely as a guess, we assume that only one farmer out of twenty had made any effort to bring running water to the house and barn and that the average distance was only 500 feet, even this modest conjecture would call for more than 1,100 miles of such piping in New York alone, and two or three times as much in the old Northeastern states. Once again, data is so entirely lacking that these figures must be termed hypothetical suppositions rather than valid estimates. In any case, boring pump logs was a very specific craft in wood, which over a period of many years must have employed the energies of a considerable number of husky craftsmen.

My own farm, among its other old-time hereditaments, possessed a pump log auger, but it had been laid aside before I came on the scene. In my boyhood, however, there was hereabouts an elderly man, one Martin Letts, who was a carpenter of sorts, and I remember hearing that he could bore pump logs. I am sure this was a job that demanded almost unlimited muscle. I am quite prepared to believe that to start boring at one end of a log and finally have the auger emerge square in the center of the other end was a feat calling for skill and calculation. Pump log augers were by no means rare tools and the Farmers' Museum at Cooperstown has several specimens. Two widely separated correspondents write of augers boring a hole one and one-eighth inches in diameter and it may well be that this became a standard size. In the beginning at least, human muscle furnished the power and I am sure that it must have been most monotonous and grueling labor. Later, toward the end of the era, some crude machinery was devised. C. F. Walters of Prospect, N.Y., quotes the memories of an old resident relative to a small shop in that community in which pump logs were turned out by the aid of two horses on a treadpower. The logs were commonly twelve or thirteen feet long, and even with horse power it took ten or fifteen minutes to bore a log.

As late as 1872, which must have been near the end of that craft, wooden water pipe was made commercially in a sawmill in Washington County, N.Y. An advertisement in the *Washington County Directory* for that year throws some light on the extent of operations and the bulky character of product:

The subscriber is prepared to fill orders for water pipe in any amount large or small. Persons ordering over sixty rods will have to send two teams. For further information write or call on the subscriber at West Hebron, N.Y. William Reid.

Pump logs of large diameter were used at Syracuse to pipe the brine in the days when boiling salt was that city's overshading industry.

The production of these makeshift conductors for bringing water from some spring to the farmstead was never a major craft, but it was important in its way, and it is interesting as symbolizing the resourceful ingenuity of the pioneer in the days when he was surrounded by what seemed unlimited abundance of timber while everything else was scarce and hard to come by.

In the year 1845, the State of New York took a census which must always be of particular interest because it is the earliest enumeration that gives us any detailed information as to what may be called the industrial and mechanical resources of the countryside. At that period we were still in the homespun age, but already it was somewhat past high noon. Men felt a desperate need for power beyond human muscle or even the strength of oxen or horse. Both gasoline and electricity were in the future. Steam power began to appear only in the fading years of the homespun age, but water power was a familiar idea brought across the sea, and from the very beginning the pioneer made use of it. Most of the small streams of the old East were harnessed at an early period. Take this very concrete example. Through the pastures of my home flows a little creek (in most of the Northeast small streams are called "cricks," and the word "brook" is rare in common speech). From the little swamp where it heads to the point where it adds itself to a larger stream is hardly more than six miles. Then there is one tiny tributary, perhaps two miles in length. It is a creek so insignificant that its upper reaches cease to run in time of drought. Just in passing, it may be said that according to tradition there was always water in the days when the dams were first built. Strung along not more than four miles of the creek's course there were to my certain knowledge ten milldams, to say nothing of a considerable reservoir. This was at the head of the two-mile tributary. The reservoir flooded perhaps twenty acres and made a water supply for the mills downstream in drought.

Suppose that some fine morning a century ago a man had decided to give himself the pleasure of exploring this little millstream from its mouth to its source. The first half mile of his walk would have brought him to a sawmill and woodworking establishment powered by an overshot wheel. The changing years have made the old pond site into a village park with trees, a little skating pool, a cobblestone bandstand, and the granite statue of a distinguished citizen, and the whole set in smooth-cut grass.

Hardly a dozen rods beyond was an old-time foundry where a water wheel took care of the small power requirements of a country foundry, including the fan which furnished the forced draft for the crude furnace where iron was melted for casting plow parts. An eighth of a mile above was a tannery, but no dam or water wheel was needed because country tanners ground their bark in a bark mill powered by one horse walking around in a circle at the end of a sweep.

Continuation of his walk a little way would bring our stroller of a century ago to another water wheel and a shop where fanning mills and the crude threshing machines of the day were made. A little farther was a towering gristmill with a huge overshot wheel twenty-four feet in diameter—a mechanical wonder built of oak and pine, the like of which is perhaps today nowhere existent in all America. It was hardly a half mile farther on to another gristmill, operating within my father's memory but gone before my time. Then a half mile or so upstream was another sawmill, and very close at hand Elmendorf's Woolen Mill, where wool was carded into spinning rolls for all the countryside and where in later years some woolen cloth was woven.

Within the next half mile were two more sawmills, and then grouped about a fork in the road a tiny but very comfortable hamlet. Here lived one who rejoiced in the pastoral name of

Millwheel

Caleb Lamb. He owned two farms, one on either side of the road. Lamb was a prominent and much-respected citizen, and so it came to pass that his community was sometimes facetiously designated as Muttonville. Right here was Betts Brothers Tannery and Hat Shop. Sam and Lou Betts were Connecticut Yankees who had come seeking their fortunes in the relatively new country hereabouts. Their business was to buy sheep pelts wherever they could be found. These were moistened, stacked up, and "sweated" so that the wool could be pulled. Then the skins were tanned and from the pulled wool felt hats were made.

Sam was what would now be called production manager. It was he who stayed at home to carry on the shop. Brother Lou, whom my father used to describe as "a great talker," was the sales force. It was his business to drive a light team of horses before a covered van stocked with hats. He roamed over the country selling hats to stores or peddling them to individuals who fell beneath the spell of his persuasive eloquence. It would seem that Betts Brothers must have been a considerable business.

Muttonville was only a short mile up the road, and there lived John Brown, who was through many years a worker on this farm. John and I were great friends and cronies in spite of a little matter of sixty years' difference in ages. We often conferred concerning things past and present, and sometimes, in a reminiscent mood, he dwelt on the departed glories of the tannery and hat shop. I specifically remember this statement that sometimes there would be employed a half-dozen journeymen. Now "journeymen" is a quaint, old-fashioned word and perfectly good English, but I think I have never heard it casually on the lips of any save John.

Sam and Lou eventually passed on and the hat shop seems to have passed along with them. A successor tried to carry on the tanner's trade, but the business could not survive the blight that overtook our rural industries following the Civil War. It was in my earliest boyhood that the once prosperous enterprise by slow degrees ceased to be. It should be added that this establishment had no pond or water wheel because the only need for mechanical power was to operate the bark mill and a horse took care of that chore.

Finally, as he pursued his walk a mile beyond the hat shop, our explorer would have come on the last of the ten millponds.

It was so near the source that even in the spring of the year the flow of water must have been insignificant. Yet today the remains of a very considerable earthen enbankment with good sized trees growing on it constitute the incontestable evidence of a venture in power development. One wonders at the hardy optimism that led a man to plan and to invest toil in such an unlikely site. Usually when a man possessed what by any stretch of imagination could be deemed a millsite, he was ill content until he had made some use of it. The foregoing is a catalogue of what might have been noted along a five-mile stretch of one little valley in the midyears of the nineteenth century. Today of all this primitive industrial development there remains only a single sawmill, still by a damsite it is true, but powered by a gasoline motor, while the once busy little stream takes its unnoted way without disturbance by any millwheel. When we lament declining rural populations and the general blight that has overtaken some of the less fortunate neighborhoods, we may remember that by no means all of this is due to farm abandonment. Oftentimes an even more compelling reason is the changing economic forces which have brought extinction of rural industrial life.

The New York State Census for 1845 makes no pretense of enumerating all the mechanical industries of the State. It concerns itself with only a few of the most important. There are just four of these that have a direct connection with rural economics. There were listed in the state 1,934 gristmills, 7,406 sawmills, 740 fulling mills and 115 clover mills. The fulling mills generally operated carding machines as part of their equipment. The clover mills were by no means universal, being restricted to those localities where the production of clover seed was an established farm practice. If these figures are added together, the total runs to more than 10,000, an impressive number. Doubtless it is true that not all of these were active in 1845, but at any rate they represented ventures that had been attempted. Once upon a time at least every one of them had a milldam and water wheel. Certain it is that all through American history until the period of the Civil War our rural industries, with an insignificant number of exceptions, were powered by water wheels built of wood and generally belonging to one of two different types.

The simplest and cheapest model was the undershot wheel, in countryside speech often called the flutter wheel. This was simply a wooden axle, fitted with wooden arm or spokes, carrying board paddles which dipped into a flume or chute where a swift jet, or stream, of water impinged upon them, and so gave the wheel rotation. Its chief recommendation was a construction so simple that it could be built by any barn carpenter. As a means of converting the energy of falling water into mechanical power, it was amazingly inefficient, but nonetheless it was the wheel that powered the primitive sawmill of the pioneer epoch. Its inefficiency was after all not such a serious fault, because the small, primitive sawmill was expected to operate for only a few weeks in spring when the melting snows or spring rains furnished plenty of water. These early mills required neither gearing nor belting nor line shafting. A wooden crank on the end of the wooden axle actuated the wooden pitman which alternately raised and lowered the wooden sash carrying the single saw blade. Each half-revolution of the wheel lifted the saw and the other half-revolution pulled it down again. The saw cut only on the downstroke and did no work on the upstroke. With the exception of a few metal gadgets, the saw blade, which was commonly about eight feet long and eight inches wide, was the only considerable piece of iron in the whole establishment.

Professor Roberts of Cornell (born 1833), searching his boyhood memories, writes of the skilled Pennsylvania carpenter who had migrated to Western New York and who "claimed that if he had a broad-ax and a narrow ax, an auger, a saw, a pair of compasses, and a two-foot rule, could build a saw-mill." There is no reason for declaring this an idle boast, assuming of course the Pennsylvanian was somehow provided with the absolutely indispensable saw blade. Outside of this one item, it was possible to make wood serve in almost every capacity. Such sawmills were amazingly primitive, as well as fundamentally ingenious, and there were literally thousands of them scattered along the little creeks of northeastern America. Not until after the middle of the nineteenth century did the familiar circular saw come to the country sawmill. With the introduction of this radically different tool came the necessity of gears and belts and precision workmanship. Perhaps Roberts' Pennsylvania carpenter might not have found himself equal to the task.

The simple, cheap, and inefficient undershot or flutter wheel might do very well in sawmills, which were expected to operate only when economy of water was not important. Gristmills, however, were in a different classification. They must, so far as possible, do business the year around. A much better type of wheel was needed, capable of using water so efficiently that even when the stream flow was small they might grind at least a few hours a day. So it was that gristmills were almost universally powered by the ponderous, slow-turning overshot. Occasionally such a wheel might be installed outside the mill in full view. This is the fashion as pictured by artists in the familiar, much idealized mill scenes. Ordinarily, however, overshot wheels were enclosed in a tight, warm building. Otherwise they would load with ice in severe weather and when this happened they were useless for power purposes. There is a mystical something about the cumbersome overshot water wheel which seems the very essence of the days that can never be again.

If the undershot wheel was wonderfully crude and cheap and inefficient, its high class relative, the overshot, was a device calling for massive, elaborate, and accurate workmanship. The famous overshot wheel at the Burden Iron Works on the Wynantskill, a tributary of the Hudson at Troy, N.Y., is said to have been the largest of its kind in America. It was forty feet in diameter and twenty feet in width, or "face." It operated almost continuously for forty-five years, often running both day and night and was finally discarded in 1898. This towering structure made only two-and-one-half revolutions a minute. Unfortunately for our story, this wheel was made mainly of iron and so cannot be called a part of the wooden age. It was, however, an advanced development of a multitude of progenitors built of wood and some of them within my knowledge had a diameter up to twenty-four feet. A good overshot, properly installed, had an efficiency fully equal to a modern turbine, with the added advantage of being able to use a small flow of water with a very high degree of economy.

As a matter of fact, the overshot wheel is even now manufactured of iron and for some purposes and situations it still remains the last word in water power. Its greatest handicap is the high initial cost and, to a lesser degree, the expense of installation. In the days when such a wheel had no competitor, it was considered that the wheel was an important part of the cost of a gristmill. I remember an old miller making a very loose statement to the effect that to build a good overshot would "take a couple of millwrights about all summer," an assertion so utterly lacking in detail that it contributes very little to the economics involved. In any case it is satisfying to remember that in the wooden age there were carpenter-millwrights who succeeded in "harnessing" the little mill stream with an efficiency that has never been greatly surpassed.

There ought to be at least passing reference to the engineering achievements of the pioneer in constructing these many thousand milldams. Of course he had no knowledge of concrete, without which there could be no modern dam building. The main dependence was an earthern embankment, with some very necessary timber and planking for the spillway. In rather special cases a substantial wall might be built of quarry stone, and this reinforced by an earthen embankment on the upstream side. Far more commonly, if any wall at all was constructed, it was thrown up of rough fieldstone, but in either case the bank of dirt was the real dependence. In the majority of instances there was no pretense of using stone. The dams were wholly of earth, hauled into place with a crude scraper or drawn to the dam site and trampled firm and watertight by the feet of the oxen that drew the carts or wagons. In some developments, the cubic yardage of dirt that was handled long before the days of power shovels and hydraulic dump trucks is most impressive. Perhaps it is true that in this soft generation we have forgotten what patient, uncomplaining physical labor through long hours may accomplish.

Building an earthen dam was a simple operation. The real problem was to provide a spillway to take care of the overflow without washing away the earth. Commonly this was done by providing a break in the earthen bank with timbered sides and with a heavy plank floor, or "apron," extending far enough downstream to prevent the overflow from undermining the dam. On the typical small stream dam, such a break would be ten or twelve feet wide and so timbered that the two-or-three-plank gates, sliding up or down in timber guides, could be raised in times of high water, or when for any reason it was necessary to drain the pond. This was commonly referred to as the "waste gate" rather than by the engineering term spillway. This arrangement for taking care of the overflow seems to have been so standardized that it must be taken as representing the accumulated experience of many builders over a long span of years. It was a general, and surely very sound practice, to set a row of willow trees on the upper side of the earthen dam just about at the water's edge. After these were well established, they filled the soil with such a dense mass of interlaced fibrous roots that water running over the crest of the embankment could do little damage. Years ago this writer had considerable firsthand experience in the repair and maintenance of a sawmill dam on the farm creek, and he can testify that the typical old-time earthen milldam required watchful attention and no small amount of maintenance.

Typically, all early water power developments were on small streams. In the days when the wooden age was in flower, men lacked both the material resources and the engineering skill to enable them to tackle the job of harnessing big creeks and young rivers. Lacking Portland cement, our present great hydroelectric developments would be impossible.

Even if the big job of erecting a dam across a large stream could not be undertaken, however, a very low dam, perhaps hardly more than a row of good-sized stones might be put across the main stream, which would deflect the flow to one bank. There it could be turned into a trench, or race, which would be carried at a grade precisely following the contour of the stream bank until the race was a satisfactory height above the bed of the stream. At a favorable point, a pond would be constructed and a wheel installed to utilize the fall, or difference in level between the diverted race and the bed of the original stream. In such installations it was necesary to put in a head gate at the upper end of the race, so that in times of high water only the desired amount would be admitted. Some variation of this plan was almost universal on the large, strong streams and it was not uncommon even in the case of small streams.

The gristmill may be deemed one of the indispensable public utilities of the pioneer community. It is true that on the very outer fringe, the far-flung skirmish line of advancing settlement, the pioneer often adopted the method of the Indian by using a mortar and pestle to pound his wheat and corn into a coarse meal. There is a widely spread tradition that the mortar was made by hollowing out a convenient hardwood stump. Sometimes the pioneer boiled his corn in weak lye until the hull would slip from it and this product after prolonged cooking became a "hulled corn," a standard product of frontier housewifery. Even so, from the very first there was the urge for breadstuffs other than hulled corn and pounded wheat, and so it was that sometimes the pioneer carried his wheat great distances to mill. There is a well supported tradition that through several years the German Palatines who first settled the Schoharie Valley, went to mill at Schenectady, a pilgrimage of thirty or more miles. Similar tales are told concerning other localities in our state. Nonetheless, as a rule, the miller followed closely the forward march of settlement. In locating, he was absolutely dependent upon water power and so it came to pass that in very many instances the accident of a stream of water, a little falls and a suitable natural site for a dam determined the location of the rural metropolis.

Unquestionably, the pioneer attached great importance to the spot where he must "go to mill." Thus about 1750 a German millwright, one Jacob Coble or Cobus built a mill on the creek, which became known, after the Dutch usage, as "Coble's Kill" ("kill" being the word for stream), and so he bequeathed his name forever not only to the creek but to the township and village as well. With the passing years the two words came to be written as one—Cobleskill. Miller Coble must have operated a very primitive plant. He had a water wheel and a pair of millstones but no bolting apparatus, so he merely ground the wheat for his patrons and each one took it home, flour and bran together, and then sifted it by hand to suit himself.

It is a safe assumption that the rude gristmills of early pioneer times made what might now be called "entire" wheat flour. The housewife expected to separate the bran from the flour and the sieves she used to accomplish this may be seen in the Farmers' Museum at Cooperstown. However, long before the memory of any one now living the bolt had been added to the equipment of the country gristmill. This was a long, cylindrical wooden frame covered with a sievelike silken fabric of different degrees of fineness of mesh. In some cases this bolt might be as much as thirty feet long, but in the little mills scattered over the countryside it was commonly twelve to sixteen feet in length and about three feet in diameter. This frame was set at a slope of about three quarters of an inch for each foot in length. The ground wheat was run into the upper end of the bolt and, as it slowly revolved, flour of different degrees of fineness was progressively sifted out and finally the bran or shorts was discharged at the lower end. The farmer-patron carried home not only the fine, white, flour which sifted through the portion of the bolt covered with the most closely woven cloth but also the next coarser and darker product which was something like the red-dog flour of modern milling. This had a name of its own—a name in daily use by any well-trained farm housewife even within my memory. It was pronounced exactly as if it were an inland waterway— canal—but it does not seem to appear in modern dictionaries except under the French term "canaille." This word also means riffraff, and probably the implication was that this coarse, dark flour was food only for the poorest of the people. As a matter of fact, the prudent mistress of her home devised ways to use it as a measure of economy and some held that for pancakes a mixture of buckwheat flour and "canal" was superior to pure buckwheat.

Almost invariably the miller received his pay for grinding not in cash but in toll, his share being every tenth bushel. This custom lingered until within my memory.

All old-time milling was done under the familiar millstones. These stones were from three feet up to as much as seven feet in diameter. The lower stone was stationary and was known as the bedder, while the upper one which revolved in contact with it was called the runner. In the early days, millstones were some-times made of native granite or such hard rock as might be available. At one time millstones were quarried in large numbers at Esopus in Ulster County, and these were spoken of as " 'sopus stones" in distinction from the burrstone (or buhrstone) imported from France. By common repute they were superior to anything found in this country and at least 150 years ago, these French stones were imported in large numbers.

The two stones were furnished with radial grooves or channels which furnished the cutting edges that served to grind the grain. These edges would become dull after a period of use and then the miller must "pick" or "dress" his stone—a rather skillful job of recutting the grooves and one for which steel tools (picks) of a very special temper were required. As a matter of fact, to temper a pick so that it would cut the wonderfully hard burrstone was a job beyond the skill of the ordinary blacksmith and so it was that certain smiths with a reputation for the work tempered the picks for a wide section of country. To my certain knowledge, more than a century ago our local miller used to send his tools to what must have seemed far-off Troy, fifty miles, for sharpening. One of the earliest memories of my life is our miller, Milton Borst, sitting on his sheepskin cushion while he patiently and skillfully picked his millstones. In these days when a very few dollars will purchase a pair of steel burrs which will grind several thousand bushels of grain before they are thrown away, it is rather surprising to remember that the old-time miller would sometimes spend six or eight days in dressing the stones. However a pair of French burrs were good for a lifetime if skillfully cared for. Perhaps it is needless to add that today the roller process for flour and the hammermill for oats and corn have made both stones and steel burrs obsolete.

The gristmill, unlike the sawmill, had to be prepared so far as possible to do business throughout the year and hence required a better type of power plant. The orthodox millwheel was the great, ponderous, slow-revolving overshot. He who has never seen (and most of us have not) the stream cascading over the huge moss-grown wheel and flinging out its diamond jets of water, has missed a spectacle which embodies the very essence of poetry and probably we will never see that lovely sight again. It is James Lane Allen, the novelist whose special province is Kentucky, who has beautifully characterized the booming rumble of the gristmill as "The earliest mechanical music of the wilderness."

The fast-changing years of the past two generations have been particularly hard on the small water-power mills, formerly scattered over the old Northeast in great numbers. The once universal custom of using flour ground from our home-grown wheat has become extinct. My sentimental regard for the old usage is wonderfully strong but I am afraid it is true that the soft, white winter wheat commonly grown in our state does not make the best quality of bread flour; certainly not unless in admixture with other varieties. Even on our farms, most of us have become users of commercially baked bread—something which I was wont to denounce as a shameful custom but one to which I have learned outwardly to conform even if inwardly protesting. The peerless buckwheat pancake has also been largely displaced by "dry" cereals sold in a pasteboard box, while home-grown corn meal disappeared from our dietary many, many years ago.

So there really is nothing left for the old-time gristmills of our state except to grind the oats, barley, and corn of the dairy farms, and now even this business has been taken over by the feed stores near the milk plants at the railroad stations. Hence there are in New York and the Northeast in general many hundreds of abandoned gristmills that will never turn a wheel again. To me, the countryside lacks a certain one-time glamour because this is so.

The gristmill and the sawmill were the earliest and most indispensable of public utilities, but they could be set up only where water power was available. Once they were established, their location inevitably became a place of resort. One industry attracts another. So it came to pass that diverse craftsmen, perhaps the tanner, the wagonmaker, and the wool carder, added themselves to the embryo community. The next logical step would be the arrival of some aggressive entrepreneur with a stock of goods for a frontier trading post. Then someday would appear a disciple of Blackstone with a law book under his arm, and now the settlers might in due and proper form make secure the titles to their lands, or, dying, might bequeath them in a last will and testament. In all likelihood, the lawyer would presently be followed by a spiritual descendant of Hippocrates. By this time there would surely be found on the scene some circuit riding, weather-beaten scout of Zion, proclaiming himself an Ambassador of the Kingdom of God, and where he trod his zeal would visibly blossom into a church with perhaps a steeple and possibly even a bell. In our earliest colonization, the church frequently preceded the school, but in due course of time some Yankee schoolmaster, product of the region where it was recognized that ripe scholarship flourished, would bend thither his footsteps, and now there would be a log schoolhouse wherein should be taught sound knowledge so that every man might cast up his own accounts and thereto set his fair signature.

Having all these utilities and institutions, the new community might be regarded as fully established. Of course in the great majority of cases these small beginnings never progressed beyond the hamlet or village stage. Now and then one became important enough to be chosen as the county seat. If so, at least a modest future was secure. It is true that Rochester might in any case have become the third city of New York State, but it is certain that the rampant growth of her early years was due to the falls of the Genesee and the water power there available. It is easy to name many towns whose origin and early prosperity were due to a stream, a waterfall, and a convenient dam site. If it be true that the man with the ax is the symbol of the possession of the land, then surely the builder of milldams and mills must be reckoned his indispensable aid and steadfast supporter.

CHAPTER XV

Shingle Shaving and Other Handicrafts in Wood

THERE is one craft in wood that was never much spoken of and which is now wholly forgotten, and yet it must be that it once had a very important place in our architectural economy. I refer to shingle shaving, or as it was sometimes dubbed, a bit facetiously, "shingle weaving." It may seem strange that the art is not included among the 374 crafts, vocations, professions, and callings listed in the occupational census of New York for 1855. Probably the reason for its absence is simply this: shingle shaving, while an important activity pursued intermittently by many workers in a multitude of places, was after all regarded as a sort of casual, on-the-side job, rather than a settled vocation. Doubtless, almost any country carpenter could easily have turned shingle shaver if the occasion required, but it is certain that they would not report their occupation as such. In southern Schoharie County (and I think this was true of all the Catskill region and everywhere in the state where men had settled on marginal land and found it hard to make a living), men would supposedly be farmers by vocation, but might eke out their scanty income by shaving shingles in bad weather. I remember James (Jimmy) Stanton, who in his old age was a resident of my home hamlet of Lawyersville. In his youth he had been a shingle shaving farmer. Near the turn of the century, when he was an old man, he still shaved a few bunches of shingles, primarily, I think, because he had available some excellent pine for the purpose and his early training reasserted itself.

Outside of the raw material, the shingle shaver's tools were of the simplest: one shaving horse or shingle horse, a simple device for clamping the shingle in place while it was being shaved; one heavy mallet, one straight frow with which to rive or split the shingle from the block, and one good drawing knife, which must be ground to an almost perfect edge so that it could roll off a wide and thick shaving. The shingle was first rived about one-third to one-half an inch thick, and then was clamped in the shaving horse. Four or five vigorous passes of the drawing knife brought it down to less than an eighth of an inch at the thin end. There was no time for hesitation nor any pretense to finish or accuracy. The worker, to make even the very scanty wage regarded as satisfactory at that period, must make not less than four hundred shingles—two bunches—per day.

Here in the Northeast the most commonly used timber was old, free-splitting white pine. Cedar was as good, possibly better, but cedar is a relatively rare tree in the Northeast. Good hemlock, while less favorably known, would make excellent shingles. On my own farm within my memory were big hemlocks which still plainly showed where they had been cut into with an ax and a chip cut out to determine if they were free-splitting enough for shingles. I suppose those left standing were the discards. Shaved shingles, split with the grain, were smooth and did not lie close. They did not water soak easily and dried out quickly. A shaved shingle lasted longer than a modern sawed shingle.

If we try to estimate the importance of the shingle industry in

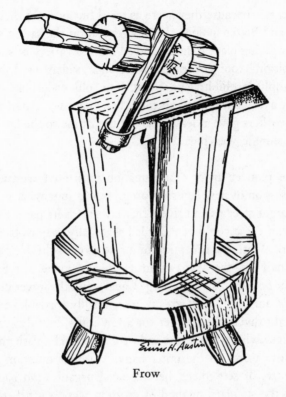

Frow

the old days, we have much more definite data to work from than in the case of pump logs. In the midyears of the past century there were in New York State about two hundred thousand farmsteads. The universal roofing material for the house, barn, and all other building was shaved wooden shingles. This reckoning takes no accounting of the scores of thousands of homes and business buildings in the towns and cities, all of which were typically roofed with the same covering. If we extend the requirements of New York State to cover the whole Northeast, shingle making becomes a really vast industry. The shingle shaver's job was secure until around the middle of the nineteenth century, when sawed shingles, made in commercial fashion, began to crowd him off the stage, not because they were as good but because they were cheaper and better looking. In this generation the sawed shingle is rapidly giving place to metal and an infinite variety of so-called "composition" roof coverings. Slate, once widely used, has become simply prohibitive in price. The old order has wholly passed away, but we may remember that there was a day when a host of workers gave at least a part of their time to the honorable craft of shingle weaving.

As the pioneer army of occupation infiltrated westward to take possession of new lands, its progress was somewhat impeded by the larger streams and rivers that lay across its path. I doubt, however, if these could be regarded as a really major obstacle to settlement. With the exception of a few rivers—the Hudson below Glens Falls, the Mohawk in its lower courses, the Susquehanna, the Seneca-Oswego (Barge Canal), and the lower Genesee—most of our New York streams were readily fordable. Some of them might divert the traveler for a few miles, but always there were places where the teams and vehicles could splash through the shallow water to the other bank. Without question in the earlier years of settlement, before social organization had been effected, the standard method of crossing was to wade in. Men and teams grew accustomed to the idea and so long as the water did not come into the wagon box everybody was satisfied. Within my easy memory before the City of New York built the dam at Gilboa and diverted the stream through the tunnel, the Schoharie Creek was a strong, bright stream. There were two fords which groups of farmers used as a short cut whenever the water was reasonably low. One of these users, James Baldwin, used to speak of "picking the creek," a phrase that had no meaning for the uninitiated. Now "picking the crick" was merely the spring chore of cleaning up the ford by disposing of any boulders that might have been rolled in by the spring floods.

Between 1780 and 1790, in piecemeal fashion, a trail was established between Catskill on the Hudson and the frontier out-post, Ithaca, in the Finger Lakes country. This path, by grace of following the valleys, managed to thread its way through the mountains by what are on the whole surprisingly easy grades. Ultimately this woodland route became the Susquehanna Turn-pike, but in popular speech it was just the Ithaca Road. It was, along with the Mohawk Turnpike and the Great Western Turn-pike, one of the three great east-west highways of the state. Eventually it was the route taken by thousands of Yankee fam-ilies, more especially Connecticut Yankees, seeking new fortunes in Southwestern New York. Along it the tide of pioneer immi-gration flowed at floodcrest for a full generation.

Leaving Catskill, there was no stream that might not be either forded or crossed on a crude bridge until the traveler reached the Susquehanna, which was a considerable river and a real obstacle to his progress. The road came down out of the Catskills via the valley of the Ouleout Creek and struck the Susquehanna just above the present village of Unadilla. Hither about the year 1784 came a Connecticut man, one Nathaniel Wattles. He must have been very much an entrepreneur because he provided both a skiff and a large flat-bottomed scow so that the homeseeker, his family, team, and household baggage, and oftentimes a little caravan of livestock, might be set across the river dry-shod and in safety. Also Wattles here established an inn where one might find lodging and entertainment and a general store where might be purchased such staples as were essential for the journey. So it was that Wat-tles' Ferry became the best known landmark of the Ithaca Road.

In 1797 Nathaniel was elected to the Assembly and while in Albany on his legislative duties, he was taken suddenly sick and died. That was a day when news, even the news of death, traveled slowly, and not until eight days later was his funeral sermon preached by the Rev. James Bacon of nearby Franklin. This ser-mon was an intimate and eulogistic memorial address. The preacher spoke concerning the deceased and how he had come into the new country along with the first wave of immigration. He had come almost without worldly goods but he had labored mightily for the good of his community. He had opened roads, he had encouraged settlement, he had ministered to the necessities of the poor and destitute, and withal he had accumulated what was, according to the modest standards of that day, a very substantial property. The regret attending his untimely death was intensified by the circumstance that he left a widow and nine children ranging in age from two to twenty-four years. It would seem that he was a strong man who deserved well of his day and of posterity.

It must be well toward a century-and-a-half since the ferry scow shuttled back and forth by the point where the Ouleout adds its waters to the Susquehanna. It was and still is a very lovely spot beside a particularly placid river. It may well be that even now if one should go there on a long summer afternoon and sit in the sun and dream, he might, across the years, hear once more the ferryman's hail and see a yoke of wide-horned Connecticut oxen before a wagon overcrowded with wife and children and household goods, come down the bank and drive out on the waiting boat; and then Wattles and his men would bend their backs to the heavy sweeps as they rowed the unwieldy craft to the further shore. There the father would pay the ferry charge from his slender store of coin and cry out the word of command to his ox team, which thereat would bow their mighty shoulders, grip the earth with their hooves, drag the heavy load up the little steep, and so take up their way to an unknown farm somewhere in "The Purchase" west of the Massachusetts Pre-emption Line.

It must have been evident, however, that fording or, when necessary, ferrying was not the right answer to the problem of getting on the other side of the stream. The construction of bridges spanning not only the small streams but also practically every river of the Northeast is a triumphant achievement of the wooden age which must not go unmentioned. At the same time there need be no lengthy dissertation on this topic because bridges, and more especially covered bridges, seem, like furniture, clocks, pottery, and early glass, to have attained the dignity of a special science. At the period when America was being settled, all the important bridges of Europe were typically of masonry. This substantial and expensive construction may have resulted from an old and established civilization with greater economic resources, but certainly in part it came to pass because of the scarcity of suitable timber. Our early building of wooden bridges was not merely the practice of old skills brought across the sea; rather it was distinctly an American development. The pioneer evolved a radically new procedure. Relatively few in number, with no accumulated resources and confronted always with the desperate task of subduing the wilderness, the newcomers had no time to undertake the slow and expensive project of building stone-arched bridges after the European models. On the other hand, they were surrounded on every side by the choicest timber in what seemed prodigal abundance, while they did not lack for husky carpenters thoroughly at home with the chalk line, the broadax, and the plumb bob.

So it was that there arose a mighty race of bridgebuilders. In New York State, the Hudson below Troy was the only stream that was beyond the skill of these self-taught engineers. It is a safe assumption that no one of them ever looked upon a set of specifications remotely resembling a modern blueprint, although it is to be supposed that some must have made certain rough sketches. Being somewhat familiar with the way in which old-time carpenters went at their job, I lean to the belief that most of the plans were carried in the back of the boss builder's head. These giants among bridgebuilders did not hesitate to bridge such rivers as the Connecticut, the Upper Hudson, the Mohawk, the Schoharie, the Susquehanna, and the Genesee, along with a multitude of lesser streams. The few examples of their work still visible excite the wonder of trained engineers accustomed to great enterprises.

With some exceptions, the important bridges over the larger streams were covered bridges. The cover of a bridge had, of course, nothing whatever to do with its structural strength, being merely a roof added to keep the framework dry. Timber, even of the kinds of wood we think of as quickly decaying, will last literally for centuries if only it can be kept bone-dry. On the other hand, if not protected from dampness, the best timber is relatively short-lived. Unless there was protection from the weather, the rain would seep into the joints and crevices, and the best bridge could not be expected to last ten years.

The early railroads until the third quarter of the nineteenth century and sometimes much later were carried on wooden bridges. In the 1890's there was a covered bridge on the railroad between Owego and Ithaca. As late as 1917 there were a number of long covered bridges on the railroads of northern Vermont. One of the most famous of wooden railroad bridges was the structure that carried the Erie Railroad across the gorge of the Genesee at Portage. Perhaps it should be referred to as a trestle rather than a bridge. It was opened in 1852 after more than two years of time, nearly two million feet of timber, and $175,000 in money had gone into its fabrication. It was 800 feet long and 234 feet above the stream—a truly wonderful example of what men could build of wood. It carried the trains of the Erie in safety for more than twenty years until one night it went up in a red flare of flame, making, as Edward Hungerford wrote, "such a pyre as Western New York had never seen before and probably never again will see."

I do not know that anyone has ever attempted an exact enumeration of the covered bridges in the timber-rich Northeast. To do so would be an impossible task. Many of them disappeared before the memory of this generation and the written records, if they exist, are not available. Some were swept away by floods, some have fallen into decay, a good many have burned, but the most potent reason for their disappearance has been the coming of the automobile and especially the heavy-duty commercial truck. Bridges satisfactory for the leisurely pace of horse-drawn vehicles could not take the pounding of multiple-ton trucks. Moreover the restricted tunnellike passageway of the covered bridge often did not afford clearance for high and bulky loads. In any case, the coming of gasoline-powered road traffic has made prematurely obsolete the many covered bridges which once spanned our rivers. Thereby the valley roads and scenery are less attractive than of old. Entering the dark archway of an echoing covered bridge gave a thrill of adventure and mystery, perhaps of romance, such as can never be experienced in whisking across the steel and concrete roadway that has replaced it.

One of the noteworthy covered bridges of New York carried the Great Western Turnpike (Route 20) across the Schoharie Creek at Esperance. It was opened in 1806, and was commonly called "The State Bridge" because it was built with state aid. This bridge took care of the tremendous horse-drawn traffic of that famous highway for 123 years, and very likely might still be serving its purpose had not the gasoline engine entered the transportation field. The southernmost bridge over the Hudson was at Troy, or more exactly a little above that city. It was a toll bridge and the nominal capitalization was $150,000, although the final cost was probably very much less. Its construction, it is said, preceded the first bridge at Albany by sixty years. This remarkable example of the bridge-builder's skill, erected in 1804, served its purpose for just a century-and-a-quarter, and made a wonderful blaze when it burned one July day in 1929. Perhaps its spectacular ending is hardly to be regretted for in any case it must soon have been razed to provide a highway for the onrushing automobile age.

There were two other bridges in New York State which ought

Hyde Hall Covered Bridge

not to go without a word of special mention. One is the distinctly museum specimen still in fine condition above the Schoharie Creek at Blenheim. It was built by a Vermonter, one Nick Powers, about 1855, and it may be taken as a perfect representative of its type, the so-called "double-barreled" bridge, which has two driveways separated from each other by a center timber truss. I have been familiar with this particular bridge for many years, but for its dimensions I adopt the measurements given by John L. Warner of Binghamton, who is in his own right an authority in this special field: "Length of clear span between abutments 210 feet. Length of truss 228 feet. Length of roof 234 feet."

This structure has been widely publicized as "the longest, single-span wooden bridge in the world," a statement which seems to have first appeared in the *Scientific American* a good many years ago. Relative to this distinction, the just-quoted Warner writes that while those who sponsored the claims of Blenheim "were undoubtedly animated by pardonable pride and commendable enthusiasm, they were skating on rather thin ice. There have been quite a number of such bridges with spans from 250 to 262 feet long. The most famous one had a span of 340 feet, and another built by the builder of the Esperance bridge had a span of 360 feet but lasted only two years. A Pennsylvania bridge, which was said to be a few feet longer than the Blenheim bridge was standing until it collapsed on September 12, 1948, and my authority believes that since that date Blenheim has been able to save face with its premature claim." Probably the safest position is to say that Blenheim has a very long covered bridge still standing in singularly fine condition, whereas all these other has-beens are gone.

In connection with the building of this bridge, there is told a local folk tale which I hope may be founded on exact fact. While the trusses were being assembled and erected, the structure was supported on block work and trestles standing in the bed of the stream. Like every community, Blenheim had its coterie of wise men, sideline superintendents, and amateur engineers who faithfully watched the progress of the job and who, after pooling their wisdom, decided that when these supports were knocked out the whole business would collapse into the creek. What really happened was just this: when, under Nick Powers' direction, the workmen drove home the last wooden pins or "tree nails," the bridge arched upward a little so the trestles stood free and clear—to the great disappointment of the experts who had assembled to witness the disaster.

Even the sketchiest reference to the wooden age bridges of eastern America must not omit the most noteworthy of them all. I believe there is no difficulty in supporting the contention that this distinction should go to "The Long Bridge" which carried the Seneca Turnpike across the stretch of wide but shallow water that forms the northern extremity of Cayuga Lake. This is to be given first place among all early bridges but not primarily because of its length. It is true that it was wonderfully long—one mile and eight rods—but there was a bridge on the lower Susquehanna said to have been even longer. Neither does the distinction of this bridge lie in the engineering difficulties involved. Indeed, its construction was wonderfully simple, its plank floor being laid on wooden stringers, which in turn rested on trestles standing in the mud not many feet below the surface. A bridge of this type called for less engineering and skilled carpentry than was necessary for the construction of the huge trusses that were to make clear spans of one or two hundred feet.

The real claim of the Long Bridge to a permanent place in history rests upon the great part it played in the settlement not of western New York but of Ohio, Indiana, and beyond. The bridge is said to have been eighteen months in building and to have cost $150,000, which was a vast expenditure for the period. A noteworthy countryside celebration was held when it was opened for traffic on September 4, 1800. The Long Bridge was not covered, but an over-the-water roadway open to the sky. Its width was twenty-two feet, which was deemed enough to permit two six-horse Conestoga wagons to pass each other. The trestles were set twenty-two feet apart, reckoned the greatest permissible span for the stringpieces. The Long Bridge passed before the days of casual snapshot photography, but happily, by grace of the facile pencil of Basil Hall, we know something of how it looked. Hall was an English artist who toured America in 1828–29 and found time to make a multitude of sketches of the contemporary scene, including the bridge and the "Grand Erie Canal," along with the "swift" packet boats which ploughed its waters both day and night at an average speed of four miles an hour.

A little east of Syracuse, two busy roads—the Mohawk Turnpike and the Great Western Turnpike—merged their westward-flowing traffic to form a new highway, officially the Seneca Turnpike, but more commonly in popular speech, the "Genesee Road." The Catskill-Ithaca Pike might be the gateway to the Southern Tier and to Southwestern New York, but the Genesee Road carried an even greater flood of homeseekers. Along the road and across the bridge through many years passed a host of families with their possessions and household goods, families who said good-by to their kin-folk in New England and to the thin and stony fields where they were bred, and who knew that in the natural course of events they would never see those fields again. When, before so many years had passed, the Purchase was full of people, the caravans still pressed on into the Western Reserve of Ohio. A little later some of them never halted until by great good fortune they came to the level reaches of the Corn Belt and the fat lands on which has arisen the most opulent farm civilization in the world.

The Long Bridge was a part of the Seneca Turnpike and that enterprise, unlike most toll roads, proved an outstanding success. It is said that for thirty-seven years it paid annual dividends of ten per cent on its capital stock. In the end, however, the Turnpike lost out. As a road it survived for many years the opening of the Erie Canal, but when finally, by the consolidation of short lines, the New York Railroad was opened to Buffalo the once great Genesee Road entered a long eclipse. No longer was the bridge crowded with traffic, and it needed a great deal in the way of maintenance and repairs. Sometime in the 1850's its day drew to a close. As late as the early years of this century a few timbers standing out of the water near its eastern terminus indicated where once the Long Bridge ran. America has bridges innumerable, bridges longer, higher, wider, and more crowded with people than this famous crossing on the Seneca Turnpike, but I doubt if there was anywhere a bridge over which passed so many entire families bound on long journeys with hearts high with hope.

Speaking of covered bridges may justify repeating a wholly

Conestoga Wagon on "Long Bridge"

undocumented and Paul Bunyan-like tale which runs thus: in 1796 which, by the way, is far enough in the past to be about the beginning of the covered bridge era, the Connecticut River was so spanned at a point north of Springfield. Along with a vast variety of lesser timbers, the plans called for twenty sticks of white pine, each to be sixty feet long and two feet square. These, of course, must be hewn with the broadax, that indispensable tool without which there could be no really important carpentry in the wooden age. To furnish timbers of such dimensions a tree must not only be of great girth and carry its diameter high up into the air, but in addition it must be "as straight as a candle." Now the most astonishing part of the story is not the dimensions of these sticks, but rather the fact that they were purchased "on the stump" in the convenient vicinity for $1.00 per tree. Today, a dollar will not begin to purchase one ten-inch board.

Assuming that somewhere in eastern America there may be remaining such pine as went into that bridge, the value of twenty trees like those would equal a king's ransom, however much that may be. Time was when men felt that our forest resources were beyond any possible need or even computation. Now we are finding out that really fine timber such as we once had in prodigal abundance has become a precious commodity. Of the once imperial forest resources of the Atlantic slope there remain only some pitiful, cutover remnants. Nonetheless, realizing something of the conditions confronting the pioneer, I have no single word of censure for him because he was an anti-conservationist. The woodland must go before the plow could come. Trees were the enemy that stood between him and smooth fields, and he who removed a tree was to be esteemed a public benefactor. It took a long time to unlearn this instinctive thinking of the pioneer years. In fairness we must neither wonder at nor blame our forebears because they wasted their once magnificent forest heritage. For them in their day and generation ruthless destruction of the woodlands by ax and fire was the only possible course. They could not conceive of a day when men would worry concerning a shrinking supply of timber.

In the days when the wooden age was in flower, the problem was to use wood for every purpose that it could possibly be made to serve. I once saw an early steam engine equipped with a wooden connecting rod. The first agricultural implements were of wood, except for a few parts where metal was absolutely indispensable. Until after 1800 that most essential of all agricultural implements, the plow, was wholly of wood except for a little wrought iron protecting the wooden share. Indeed certain large makers of plows retained the wooden plow beam until recent years. Harrows made wholly of wood, except for the iron teeth, are within easy memory. The first mowing machine I can remember, the Buckeye, had a very substantial wooden frame to which were bolted the bearings that carried the shafting and gears. One of the minor, and yet important, devices of the era was the use of shoe pegs, supposedly because the old-time cobbler found it easier to whittle a wooden shoe peg than to buy an iron nail. Evidently, the method was not too bad; in making heavy farm boots, the shoe peg held the field long after metal pins became readily available.

This is not the place to make anything more than a passing reference to wooden clocks. Books have been written concerning them. Some wooden clocks were made in Europe but it remained for the resourceful Connecticut Yankees to turn a small beginning into a really major industry. Rolled brass was not only expensive but hard to come by. Yankee mechanics turned to wood, the material they knew best. A serviceable wooden clock could be made and sold for half the price of those made of brass. If kept in a dry place it would run for, as some have, a hundred years and more. The great days of wooden clocks fell in the first third of the nineteenth century. Today, when almost everyone below the age of fifty wears a wrist watch, and when the home without electric clocks is regarded as lacking in modern conveniences, it is hard for us to realize that until after the turn of this century, watches and clocks were uncommon and confined mainly to families of acknowledged wealth and position. Lacking such man-made contrivances, countrymen became amateur astronomers and regulated their lives according to the position of the heavenly bodies. By dint of much experience, a man in the fields could make a close calculation as to when it was dinner time. At evening he needed but to lift his eyes to the western horizon.

This dependence upon the course of the sun entered into the casual everyday folk speech of the countryside, as recorded in certain court testimony of the period. On the ninth of October, 1818, a Schoharie County farm was the scene of a lurid murder. This is no place to relate the details of that crime beyond this single circumstance: there were three neighbor men who were present at the farm the afternoon just preceding the crime. These three talked with both the murderer and his victim in friendly fashion. Then they took their departure and, with the exception of the killer, they were the last persons ever to see the victim alive. Naturally they became important witnesses at the trial. One of them, on examination, being asked as to the time when they left the farm, made answer, not in the speech of today, that it was about half-past three, but instead that he "judged it lacked about an hour-and-a-half of sundown." Thus did that countryman's language betray the thought habits of a lifetime.

Finally, in cataloguing the uses of wood in our economy, we must on no account forget that for the first two centuries of the white man in America wood was the practically universal fuel, and that even today in a great number of rural homes it occupies a prime place both in keeping warm and in cooking victuals. On the typical Northeastern farm "getting up the wood" still remains a major winter industry. It is true that there are, here and there, farmers who have so far forgotten their ancestral training as to depend on oil or coal for heat and electricity for cooking, but most of us retain at least a wood fired kitchen stove, and not infrequently you will meet a farmer who can assure you with a note of pride that he has still to purchase his first ton of coal. It may be that we farm dwellers must be reckoned a dour and obstinate breed.

It is said that veins of Pennsylvania anthracite were opened as early as 1768, but with only wagon transportation, retailing this coal remained a purely local industry. Even as late as 1820 there was an almost negligible production of only 365 tons. Some small shipments reached Philadelphia by floating down the Lehigh River into the Delaware. The completing, however, in 1828 of the Delaware & Hudson Canal offered dependable and cheap transportation between Honesdale in the coal country and Rondout on the Hudson. For forty years thereafter this waterway enjoyed a sunburst of prosperity, until the growing railroad systems proved its undoing. It was coal brought to tidewater by this canal which began to displace wood as a fuel for cities.

Any general or widespread use of coal as fuel though, awaited the development of a comprehensive railroad system and this was pretty sketchy until after the Civil War. In fact, the locomotives of the earlier railroads used wood for fuel. In the 1850's and into the 1860's, the New York Central engines burned wood. St. Johnsville in the Mohawk Valley was one of the principal fueling stations. As a small boy I was taken there rather frequently because it was the home of my mother's people. Coal had displaced wood before my earliest memory, but my father's recollections ran back to the days when an almost unbelievable quantity of four-foot cordwood was hauled in from all the surrounding countryside. Perhaps almost no one remains who can really remember, but it is said that when a wood burning locomotive was pulling a heavy train up a grade on a dark night, the stream of sparks that poured from the stack suggested a giant Roman candle in a state of permanent eruption.

Following the success of Robert Fulton's *Clermont*, steamboating on the Hudson entered a proud period which endured for a century. Today only a pitiful remnant of that once great fleet remains. The reason that makes it here permissible to recall the halcyon days of the white-and-gilded craft of the American Rhine is the circumstance that during all the early and midyears of that period it was wood and not coal which stoked the roaring fires beneath their boilers. These fires had always an insatiable appetite for more wood and eventually both sides of the river were largely denuded of timber. It is said that the scrubby pine which once covered large areas of the sandy ridge between Albany and Schenectady was cut in wholesale fashion and drawn to Albany for "steamboat wood."

The big stone fireplace which was the central heating plant of the log house demanded plenty of wood, but, after all, it took only a small fraction of the amount available. As soon as sawmills were built and some sort of market for lumber established, the settler would save some of the choicest logs, especially the pine which has always been esteemed above any other timber. James Baldwin, the before-mentioned graybeard with whom my boyhood memories are linked, told me that in pre-Civil War days he drew pine lumber from southern Schoharie County to tidewater on the Hudson. It was a long, hard, two-days' journey—more than sixty miles for the round trip. Then he added: "We never took a board if it had one little knot or even a 'gum spot.' " If he told me for how much timber sold, I have forgotten. In any case, the price represented only scanty pay for the toil that had entered into its production. The raw material on the stump was still almost without value. When village living developed to the point where there was a market for firewood, the accepted unit of measurement was the four-foot cord, or 128 cubic feet. In practice, this meant that the sticks were cut four feet long and laid in a pile eight feet in length and four feet high. As early as the eighteenth century, the Hudson River proprietors were writing into the perpetual farm leases which were paid "in kind" and not in money, not only the stipulated bushels of "sound, sweet, merchantable winter wheat" and the "four fat Fowls," but also the annual tribute of two cords of wood to be delivered at the manor house.

The comparative heating value of different kinds of wood is measured almost exactly by the weight per cubic foot when thoroughly dried or, as the countryman says, "seasoned." So it is that in popular thinking the so-called "hard" woods are given a high rank because of their greater specific gravity. The forestry experts have prepared tables giving the relative fuel value of the different species. Hickory is accorded first place, closely followed by sugar-maple, beech, birch, and oak, while pine and basswood are at the bottom of the list. If wood were sold by weight instead of bulk, it would be a somewhat different story.

A cord of hardwood will weigh from two to two-and-a-half tons, and a good team could draw it on a sleigh provided only that the grades were not too steep. In practice, a half-cord was the more usual load. Experienced opinion and scientific determination agree that a cord of seasoned hardwood has approximately the same heating value as a ton of coal. As late as 1850, the output of the Pennsylvania anthracite mines was less than one million tons annually—a clear indication that wood was still the chief domestic fuel of America. Today, outside of the farm kitchen, wood is recognized as a luxury fuel and a living room with a fireplace and crackling wood fire is regarded as the very symbol of gracious living. There is no other heat so conducive to dreams and memories as an open fire.

Cord of Wood

CHAPTER XVI

Wooden Age Occupations

THERE were three industries of major importance in the pioneer period and they are most logically considered at this time. The first two were possible only while we had abundant forest resources and may now be deemed extinct. While the third still lingers it is only a shadow of what it once was. These three crafts are the making of charcoal, of potash, and of maple syrup.

Charcoal should be considered first because it is really a further discussion of wood as a fuel. At one time charcoal was a universal staple of frontier commerce. It was the sole fuel of the indispensable blacksmith. It was the fuel with which the tinsmith, sometimes called the tin knocker, heated his soldering iron. Charcoal was the fuel that in unbelievable amounts entered into our early iron industry. It is said that it took about two-and-one-half tons of charcoal for each ton of pig iron. The first successful iron works in America (Virginia had tried it earlier and failed) was established near Lynn, Massachusetts, in 1643, and it ultimately attained a production of "seven or eight tons a week." It may well be that the first demand for conservation in America was when, forty years later, the local citizens protested that the demand for charcoal threatened to denude the countryside of all timber. For the first hundred years of iron making in America, it was Massachusetts rather than Pennsylvania which was the center of the industry. Iron ore is widely scattered in many localities, and eventually there arose an astonishing number of ventures in smelting it. Used by every blacksmith and in all sorts of metallurgical operations, charcoal had an important place in commerce and was one of the few commodities with a dependable cash market. Many a settler eked out his always scanty income by turning some of his woodland into charcoal. This was common practice as late as 1860, and even a generation later there might here and there be found a man still carrying on.

More than twenty-five years ago there was still among us now and then an ancient man whose memory ran back to the days and methods of charcoal burning and potash making. Through appeals and inquiries in the columns of *The American Agriculturist* I had the names of a number of such men and correspondence with several. The most exact and satisfying data regarding charcoal was from a worthy in Oswego County. In quite specific fashion he wrote me to this effect: wood for charcoal burning was commonly cut about four feet long. Any kind of wood might be used, but elm was considered especially desirable. The man who laid the kiln began with a pile of light, dry kindling wood in the center. Around this kindling, the wood was set on end leaning toward the center, and this was continued until there was a circle of wood about twenty feet in diameter. On top of this lower pile another pile was constructed, the wood being set on end as before. A pit generally contained twenty-five to thirty cords of four-foot wood. When complete, the pile was thickly covered with earth and sod, and in general shape might have suggested an old-fashioned straw beehive. Openings were left in the earthen covering so as to give some draught at the beginning. When all was ready the pile was fired by thrusting burning poles into it, and as soon as it was well started all openings were closed with earth.

To burn a pit satisfactorily required no small degree of skill, judgment, and vigilant care. If a blaze threatened to break through at any point, the place must immediately be covered with fresh earth. The danger of the slow, smouldering combustion turning into a swift conflagration was so great that the pit must be watched day and night for nearly two weeks. So it was that the burner built a little cabin where he might be sheltered and possibly catch cat naps while he watched his fire. Sometimes two pits were laid close together and the cabin built between so that a man might guard two fires instead of one. My informant wrote me that a cord of four-foot wood commonly made about thirty bushels of charcoal. It would seem that, theoretically at least, it ought to make more, but these are the figures given me. When finished, the product was drawn several miles to Oswego City, and sold for fourteen cents per bushel. In other words, after all his labor and care, he received, say $4.20, for a cord of wood—an indication of how our fathers toiled for a little cash money.

It would seem entirely germane to the topic if I repeat a family folk tale of which my own great-grandfather was the hero, or, more exactly perhaps, the victim. In a previous chapter, I have set down his somewhat remarkable last will and testament which, among other things, reveals his fatherly love and thoughtful provision for his unwed daughter, Sarah. Many years before, in his young manhood, he experienced a misadventure which eventually grew into a permanent part of the family annals. Great-grandfather had burned a pit of charcoal which he planned to sell in Albany. At evening the wagon, topped with a high, slatted, charcoal rack, was loaded so that a very early start might be made next morning. That night he slept the sound sleep of the weary toiler, but at first dawn he arose and harnessed his team for the trip. When he came to the clearing where the load had stood, great-grandfather was dumbfounded to find the irons of the wagon and nothing more. What had happened was after all no mystery. Somewhere in the load there remained a tiny spark of fire and when the air reached it there was a general conflagration. For a pioneer farmer just getting established, the event was no small catastrophe, involving not only the loss of the load but, much more serious, his best, and very likely, his only wagon.

Charcoal Burner

In the age of homespun our present-day chemical industry played only a minor part, but even then potash was eagerly sought after and held a large place in world commerce. It was used in glass and soap making and in dyeing, but most of all in the scouring of wool, meaning thereby, cleansing the wool of the yolk, the natural gummy secretions which ordinarily account for more than half the weight of the fleece as shorn. England had long been the world capital of the textile trade and the preparation of the wool called for large amounts of potash. Before the potash mines of Germany were opened up in the 1850's, the commercial supply of this indispensable commodity was derived from leaching wood ashes, mainly in the United States, Canada, and Russia. Dr. U. P. Hedrick writes me that in the heyday of the potash trade with Europe, America's export of the product went abroad mainly through the port of Montreal rather than New York. This would be a feasible route after the construction of the Champlain Canal, and, in fact, was always the most practical shipping point for the northern counties of the state.

On that great day in October, 1825, when the Erie Canal was formally opened to traffic throughout its length, one Jeptha Simms was a clerk in a general store at Fort Plain on the Mohawk. With the passing years he became an antiquarian and the local historian of the Mohawk and Schoharie Valleys, and he wrote concerning many things. More than fifty years after the event he set down his recollections of the great celebration which marked the completion of the canal. As a merchant in a store beside the towpath, he was familiar with the history of the early and palmy days of that greater waterway. He remembered, too, the bateaux and the Durham boats that even before the building of the canal plied the Mohawk River in large numbers, and he records specifically that the three most common items of east-bound freight were wheat, potash, and whiskey.

In any case, the manufacture of potash was, a century or more ago, one of the important industries of our state. In 1845 (I must continually refer to this date because it is that of our earliest statistics) there were in New York State 738 asheries and the value of their product is given as almost a million dollars ($909,194), which was an imposing sum of money in those years. Ten years later the number had dropped to only sixty-eight. Evidently it was a rapidly dying industry. As late as 1865 there were reported 54 asheries, although it may be assumed that some of these were dormant rather than going enterprises.

For some firsthand information regarding the making of potash I am indebted to Lot Hall of Gouverneur, St. Lawrence County, a farmer of rare intelligence and exact memories, whose friendship I cherished through many years. By vocation he was a dairyman but he was so much the student that he became a self-taught botanist of no mean attainments in his special field, the native grasses of his locale. He survived to a rather remarkable age, but it is now a fairly long span of years since he died. He was born in 1844, a date still well within the pioneer period in that region, and as a boy he was familiar with the primitive technique of boiling potash, an art, by the way, which seems to have been more highly developed and to have lingered longer in St. Lawrence County than in any other part of the state. He wrote me at length and with considerable detail concerning the practice of this one-time-important craft.

The trees were chopped down, cut into convenient lengths, hauled together, and rolled into piles along the limbs and brush, and then the whole heap was burned. As soon as the fire had died, the ashes were gathered and leached, and the resultant lye boiled in big potash kettles shaped like a half-eggshell. The bottom of the kettle was cast thick so as to endure the strain of evaporating the lye down to a solid mass. The resultant cake was known as "black salts," and was dark in color through carbon and other impurities. This black salts was the form in which potash was sold by the farmer. The usual price was about $3.00 per hundred pounds, and Mr. Hall stated that this was sometimes the only cash money the pioneer saw. At this same period, butter was fourteen cents a pound but this was "in trade," as the phrase was, and not in cash. Making black salts was a work not without its risks and uncertainties. Sometimes after the burning of a huge log pile, which represented a week of hard work, a sudden heavy shower would leach the ashes before they could be gathered, thus entailing a total loss.

The asheries, which purchased the black salts from the producers, were establishments thickly scattered over some parts of the state. At these asheries were brick kilns in which the crude salts were burned at a high temperature, consuming the carbon, and fusing the mass into a bluish-white and much purer product, then known as "pearl ash." Not all farmers did even their own leaching and boiling. Some asheries maintained teams which went around the countryside collecting the ashes and paying eight cents a bushel for them—a rather surprising price in that day when a dollar seemed a very considerable sum of money. I do not wonder that St. Lawrence County had ninety-seven and Cattaraugus thirty-three asheries. It would seem that making potash was one of the most universal and remunerative of pioneer activities, but at the same time it depended upon the exploitation of a natural resource that was definitely limited.

As told by Lot Hall, there was a wide difference in the potash value of the various species of trees. By common consent the very best was water elm. He said that a single large tree of this species might be expected to yield as much as 200 pounds of black salts and the ashes were so rich that in gathering them it was not unusual to find solid masses of fused potash "as big as a large potato" and so pure that they would be thrown directly into the boiling kettle. Next in order of richness he rated black ash, maple, basswood, hickory, and beech, adding that all the evergreens were so poor in potash that they were not deemed worth cutting and burning. These rather loose generalizations were the opinions of a very old man remembering his youth, and possibly they would not in all details endure rigid scientific examination.

As a further contribution to our knowledge of this early industry, there is this singularly informative advertisement as it appears in the *Schoharie Republican*, issue of March 31, 1824:

ASHES WANTED:

Fourteen cents per bushel will be paid by the subscriber in goods at the lowest cash price at his store in Middleburgh for any quantity of good clean house ashes, delivered at his ashery from this time until the first day of May next. And to those who cross the Middleburgh bridge with ashes for him, one shilling extra in cash will be paid for every load of fifteen bushels or upwards.

Middleburgh, Dec. 8, 1823   CHARLES WATSON

This announcement adds to our understanding of that period on four distinct points. First, it indicates something of the going price for ashes in that locality and probably in eastern New York in the year 1823. Secondly, the stipulation "good, clean house ashes" is in contradistinction to the ashes gathered up where log heaps had been burned and which were likely to be of less value because of the danger that they had been partially leached or were mixed with earth. Thirdly, the phrase "to be paid by the subscriber in goods at the lowest cash price" speaks of the days when rural transactions were very commonly settled by barter rather than in cash, an idea that lingered among old-time country merchants even within quite recent memory. Fourthly, the concession of "an extra shilling" for loads that must cross the bridge over the Schoharie Creek at Middleburgh gives us at least some idea of the toll bridge charge at this date. Just what was the rate at this time we do not know, but judging from the tariff on comparable bridges at this period it seems very probable that the charge was a shilling for a wagon and two horses. In any case, note that while the pay for the ashes must be taken "in trade" or "in kind," storekeeper Watson recognized that the toll must be paid in cash and so offered the "extra shilling."

Charcoal burning and making potash from wood ashes are two industries now not only extinct but almost forgotten as well. Few among us will know of them even by name; fewer still will comprehend anything of the aid and comfort they gave the pioneer in those first hard years in the new country. They were noteworthy in their time and they deserve at least this word of remembrance.

Potash boiling, charcoal burning, and maple sugar making were three activities that might be undertaken just as soon as the pioneer made a beginning at chopping out his farm. The first two could be carried on almost anywhere if wood were available; the kind of timber was not too important. Both of them have been practically extinct since Civil War times. The supply of raw material is no longer available; the once eager market has disappeared. Maple syrup and maple sugar making, however, are still with us although the industry is now only a shadow of what it once was. The first effort to gather statistics relative to the industry was the federal census of 1860 when there was reported the equivalent of nearly six and two-thirds million gallons of syrup. Apparently the business had even then already passed its zenith because this early production still remains an all time high. Our present country-wide production seems to be somewhere around forty per cent of that and varies sharply with the weather conditions. New York and Vermont each make about one-third of the total, and the other third is contributed by half a dozen different states. New York still taps between two and three million trees, which even in these days of decimated forests represents only a minor fraction of the possible number. It would seem that the business is on its way out, but probably it will never really come to an end. The product is so incomparably delicious that there will always be an assured market at a relatively high price.

According to a universally accepted tradition, the art of making sugar from the maple was a discovery of the American Indians. There is in European literature no reference to similar work and competent botanists declared that there is no European tree yielding a comparable sweet sap. Granting that the Indians were the original processors, the white arrivals promptly took over. How rapidly the settlers and the aborigines traded information with each other! The first-comers to the shores of the eastern seaboard immediately adopted that strange grain, maize, and for a name took the old generic term "corn." In like manner they made a place for the utterly unknown beans, squash, pumpkin, artichoke and tobacco—all New World contributions to the White Man's agriculture. So, too, the Indians were equally adaptive, carrying apple seeds and peach pits hundreds of miles beyond the advancing frontier.

The Iroquois, of course, had no auger. The best he could do was to take his stone ax and cut a groove through the bark of the maple, extending obliquely downward and perhaps a foot long. This groove collected the sap and carried it to a point at the lower end of the cut. Inasmuch as such a channel opened up a large amount of sap-bearing wood, it is reasonable to believe that a freshly scored tree might yield sap very freely. At the bottom of the groove, a gash was cut in the bark, and into this was driven a thin splint or sheet of wood not unlike a short shingle, to conduct the sap into the container. This method of tapping, which was really partial girdling, was ruinously destructive to the well-being and long life of the tree, but this fact was not at all important when a few scattered Indians were surrounded by unnumbered thousands of maples. There is no doubt that the first tapping done by the whites followed the Indian method, but the use of an auger to bore a hole and a wooden spile or spout soon became standard practice. To catch the dripping sap the Iroquois may have used bowls or troughs carved, dug, or scooped out of some soft, easily-worked wood. In our Northeastern forests, basswood is almost certainly the timber best suited for this purpose. On the other hand, the Iroquois may have used earthen vessels or quite possibly dishes made of bark.

There remains the question as to how these Indian sugar makers boiled down the sap. Here we have suppositions, traditions, and folk tales rather than specific knowledge. Students of pre-

Indian Sap Trough

European Indian life believe they used earthen pots and Carl E. Guthe, director of the New York State Museum, tells me that such pots have been found blackened by fire on the outside, which fact may be taken as conclusive evidence that they were used as cooking vessels. The Iroquois were past masters in the use of bark, and it would appear wholly reasonable that they made bark troughs for boiling sap. Just as it is perfectly easy to boil water in a plate made of reasonably waterproof paper, so it would be quite possible to use a container made of bark so long as care was taken to keep the liquid always as high as the part touched by the fire. The outside would, of course, blacken and char, but it could never burn through to the liquid within.

There remains yet another myth which seems to have a perennial vitality. It is to the effect that the aborigines boiled down their

Tapping a Maple

sap by the expedient of throwing hot stones into it until the job was complete. In spite of the fact that this tale has the authority of frequent repetition, I dismiss it as utterly beyond credence. It would take a considerable pile of very hot stones to evaporate even a single gallon of sap. The whole idea would be inefficiency raised to the nth power. Then the plan would surely fail after evaporation had reached a point anywhere near sugar. There remains the further fatal defect that after hot stones had been thrown in, fished out and reheated so many times the end product would have been so largely potash lye that its acridity would have been beyond the stomach of even "poor Lo." I am sorry to express such frank disbelief in a time-honored legend, but I want to retain my intellectual integrity.

In the days when money was woefully scarce and self-sufficiency the prime motif in farming, maple sugar was distinctly a home acres product and the settler made much use of it. In the earliest years as vessels to catch the dripping sap, he sometimes used troughs dug out of a short piece of basswood log. A perfectly authentic specimen of such a container is now in the Farmers' Museum at Cooperstown. This expedient, however, was at best a rather pitiful makeshift. Very soon the cooper took over, turning out rough but efficient sap buckets literally by the millions. To boil the sap, iron kettles seem to have been used from the first, for kettles came with the colonists on the first ship. In the more primitive operations the kettles were merely hung over an open fire. If the maker sought a somewhat higher efficiency, he laid around the kettles a so-called "stone arch" which retained the heat where it was needed, saved fuel and minimized the frequency of sparks falling back into the boiling sap. If his operations were at all extensive, he built himself a "sap-house," where he and his work might be sheltered from the weather.

The passing years have altered this. Today, the industry has undergone the same sweeping changes that have come to other farm activities. Wooden spiles and buckets have been replaced by metal spouts and tin or galvanized iron buckets, while the big kettles have made way for efficient evaporators made of copper with corrugated bottoms to increase the area exposed to the fire. In these years, the expert in charge will determine when his product has reached the proper degree of concentration not by the way in which it froths and bubbles in the kettle, nor yet by the test of hairing when a few drops are poured from a spoon, but by noting the temperature of the boiling pot with proper allowance for varying barometric pressure. In a word, exact laboratory methods have replaced the time-honored rule of thumb.

Sap will run at any time after the leaves fall, provided only that the nights are cold enough to freeze sharply and the days warm enough to thaw freely. Such days may fall now and then in February, but in practice the sap season usually lies between mid-

March and mid-April, with rather wide variations from this rule. The yield of sap varies widely in different years, depending primarily on weather conditions. There is the surprising fact that sap will run well only with northerly or westerly winds. Temperature conditions may seem ideal but if there is east or south wind, the flow will dry up—a phenomenon for which the plant pathologist has no adequate explanation. Different trees vary widely in the sugar-content of their sap, and this may range from two per cent to five per cent. The yield per tree varies greatly, but in a good sap bush with proper attention, there may be expected, one year with another, say, three pounds of sugar or three pints of syrup for each bucket hung.

Sugaring time has long been (perhaps still is) something of a rural spring festival. When sap is running freely, it is often necessary to boil all day and far into the night. A snug sap house with a roaring fire and a pleasant aroma of boiling sap easily becomes a place of evening resort, more especially for the young folks. Eggs boiled in the foaming sap furnish the simplest of banquets. Syrup cooked down to the point where it cools to sugar and then poured on snow becomes the "jack-wax" of blessed memory; or, if preferred, each guest may have some in a saucer and when stirred with a spoon it will become "maple honey," which even now remains the most delicious of confections. Those of us who grew up in communities where sugar making was in many cases a standard farm activity, can hardly fail to have nostalgic recollections.

Somewhere in this book there should be a discussion emphasizing the important role of the honeybee in the pioneer economy. Most logically, I think this is best done just at this point. Sugar and syrup from the farm maples and honey from their own bees were the housewife's answer to what must have been a chronic sugar shortage. Remembering the high price of such sugar as might be found in the stores and the very small amount of cash available on the frontier, it is evident that the sap bush and the bee yard had an imperative importance which has now largely disappeared.

Our familiar honeybees are natives of the Old World and were unknown in America until the white man brought them. It is said that they arrived in New England as early as 1638. The life habits of a colony of bees is such that their survival on the long voyage in the slow ships of that period must have been an uncertain venture. Without doubt, however, bees were considered almost indispensable and it is altogether likely that they were brought to other parts of the Atlantic seaboard. One Benton traveled in the colony of New York in 1670, and he has left behind this rather idyllic observation: "You will scarce see a house but the south side is begirt with hives of bees which increase after an incredible manner." Another traveler, Thomas, writing of Philadelphia in 1698, says: "Bees thrive and multiply exceedingly. The Swedes often get great stores of them in the woods where they are free for anybody."

It may well be that there is a measure of truth in the belief that bees prosper best on the fringe of civilization. Bees are now subject to two very serious diseases, the American and the European foul brood. Both diseases are characterized by the fact that the larvae, that is, the immature bees, die in the cells before hatching out as mature insects. Presumably we did not have these troubles in the early days, so it may well be that the risks and uncertainties of apiculture were less than now. In any case, once introduced into America, bees prospered amazingly in their new environment. Swarms were always escaping to the woods; a hollow tree furnished an almost ideal home. As a matter of record, the honeybee crossed the Allegheny Mountains far in advance of the first settler. Remote Indian tribes, hundreds of miles from the seaboard, recognized an insect strange to their experience and named it "the white man's stinging fly." It is an old-time whimsy that absconding swarms always fly westerly, an idea doubtless without any real foundation. The direction of their flight is straight toward the hollow tree which advanced scouts have already selected as a promising place to set up housekeeping. America's bee husbandry was built around the black (more correctly brown) or German bee, the race found in northern Europe. About 1860, the yellowish, sometimes almost golden, Italian bee was introduced and has been very widely disseminated; it has, however, never fully displaced the German bee and today most of our stock is a mixture of the two races.

A remarkable amount of folklore has from the earliest times attached itself to bees and some of it has lingered on to a surprisingly recent date. Within my memory the issuing swarm was promptly greeted with the beating of tin pans and the ringing of bells, supposed to lead them to cluster in the immediate vicinity instead of flying away to the woods. Another belief which must have had its roots in some far-off century, decreed that if there was a death in the family the bees must be "told." Otherwise they would leave the hives and fly away to the woods. This "telling" was done by fastening a little shred of black cloth to every hive. To my knowledge, this bit of medieval ritual was still enacted at a period almost or quite within my memory, and, moreover, not four miles away, by a family who have been with us since the beginning of settlement in this region and who were recognized as people of substance and standing. Incidentally, it was the women of the house who took upon themselves this melancholy duty.

Beekeeping a century or more ago was crude and primitive to a degree almost beyond belief. The type of hive most commonly used was merely a section of a hollow log, perhaps two feet long, with a board nailed over one end for a cover. Such was called a bee gum, and might now and then be found in use until a period within my earliest memory. Just as a place for bees to live, breed, and swarm bee gums have never been surpassed. The thick walls afford good protection against winter cold, while the ideal shape for a bee nest is circular rather than rectangular. The oft-pictured conical hive, built up of bands of twisted straw, may be regarded as the symbolical hive of artists. I do not know how widely this form was ever used in America, but it is certain that it was sometimes found. At the Centennial Exposition at Philadelphia in 1876, Captain Hetherington of Cherry Valley, N.Y., made a noteworthy exhibition illustrating apiculture as it then existed. For historical reasons he needed a straw hive, but had difficulty in locating such a specimen; however, the need was finally supplied by the Snyder family of Hyndesville, people whom I came to know in later years. The family had been hereditary bee masters for a great while and well into this century they were still keeping bees, but not in straw hives. An example of this rare

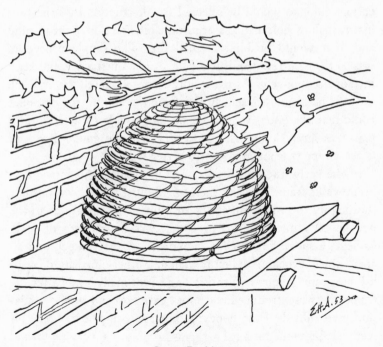

Straw Beehive

construction may be seen in the Farmers' Museum at Coopers-town.

Ultimately, most people settled down to the use of the box hive, which was nothing more than a box made of inch boards. Commonly, it was about a foot square and a foot-and-a-half deep, but the exact dimensions were not standard or important. There might be some holes bored in the top and arrangements so that small "supers" or "caps" for the storage of honey could be added. Sixty or seventy years ago, when I was getting my initial experience in apiculture, the box hive was still very common, and my first love was for bees in such a home.

Until less than a century ago, harvesting the honey crop was a crude and cruel operation. In the autumn, when the hives were heavy with honey stored up against the needs of the long winter, certain colonies would be selected for destruction by brimstoning. A shallow hole, called a brimstone pit, was dug in the ground and in it would be built a small fire. The simple procedure was to throw a few handfuls of brimstone on the fire and set a hive of bees above it. As the deadly fumes mounted through the hive the bees almost instantly fell from the combs; the despoiler could take the booty they were no longer able to defend. This plan was surely killing the goose that laid the golden egg, but it was a time when men knew no better way.

In the early 1850's, Lorenzo Loraine Langstroth introduced his movable frame hive and then for the first time it was possible thoroughly to inspect the inside of a hive without injury to the occupants. Langstroth (1810–1895) was a graduate of Yale College, a successful teacher, a minister of the Gospel, and an apiarist who probably did more than any other one man to establish beekeeping upon a scientific basis. Not many years after the frame hive was introduced, the honey extractor and comb foundation were added to the honey producer's equipment. Once his feet were set in the forward path, his progress was very rapid, and by 1875 the business had fallen into what seems its present pattern.

Unfortunately there are no early census figures but it seems certain that apiculture today is only a shadow of what once it was. More than 250 years ago, Benton observed, "Scarce a house but that the south side was begirt with hives of bees," and within my memory a half-dozen or more hives (often called "skips") in some sheltered spot not too far from the kitchen door were a common feature of the countryside. Like so many other vocations, the industry has passed into the hands of a comparatively few large producers who are specialists. Holding nostalgic memories across a good many years, I feel, with a quiet pain of regret, that "the glory hath departed."

It is an old idea that bees thrive best in the new country. Hence the ancient quatrain:

> Ding dong of bell and choral swell
> Deter the bee from industry
> But hoot of owl and wolf's long howl
> Incite to moil and constant toil.

CHAPTER XVII

The Sad Story of Silk Production and the Success of Tanning

JUST for the record, there should be some brief mention of the many attempts over several centuries to produce silk in America. These began in the early years of the Jamestown colony and persisted intermittently in many different regions until the period of the Civil War. Occasionally, as in Georgia, Kentucky, and the Carolinas, these efforts attained some small measure of success. Silk culture always enjoyed the advantages of governmental approval and support. In 1826 the federal government authorized the publication of a *Manual of Silk Culture*. The book ran to more than 220 pages, and it was ordered that six thousand copies be printed. I assume that this was one of the very first forerunners of that innumerable flock of bulletins which have originated in Washington and which through the years have offered free advice on almost every imaginable topic. Many state governments, as well, tried to give aid to this infant industry which, unfortunately, never lived to grow up.

Evidence of the popular and official interest in silk culture is found in the fact that the New York State Census for 1845 selected, apparently in rather arbitrary fashion, just twenty-one industries to be examined as to their activities and economic importance. One of these was silk culture and the reported progress could hardly have given much satisfaction to its proponents. Some production of raw silk was reported from twenty-nine different counties, indicating rather widespread interest, but the sum total of 1,439 pounds for the entire state must be regarded as insignificant. Monroe was the banner county of the state, producing 283 pounds or about one-fifth of the whole. Some hardy and optimistic experimentalist of Chenango County reported a yield of one pound of silk, which represented the entire production of his county. It is a bit absurd that silk culture, which at best was hardly more than an hypothetical industry, should be so carefully examined and catalogued, while beekeeping, a distinctly going and everywhere present activity with a relatively large total production, gets not even mention.

The golden age (if high hopes and boundless enthusiasm deserve that term) of silk culture fell in the second quarter of the nineteenth century. At one time there were in existence four monthly magazines devoted to this industry, two of them published in New York City. It is hardly possible within the range of written language to hold forth more rosy prospects than are offered the fortunate souls who will cast in their lot with those who give their energies to the production of silk. One contributor writes in a burst of lyrical enthusiasm, "But the object which we, as Americans, should have in view is not to be bounded by the amount of our own consumption but we have a far more brilliant consummation before us which is to supply the whole continent of Europe with the raw material, whose annual consumption cannot amount to less than seventy-five to one hundred millions of dollars." The author proceeds further to let his imagination and his taste for statistics run wild. He estimates that this vast quantity of silk will call for 500 million mulberry trees to furnish the leaves which, as everyone knows, are the favorite provender of this wonderful insect, the silkworm. It would seem that all other agricultural activities were to be dwarfed by comparison.

If silk and silkworms became a vast enthusiasm, then the propagation and sale of cuttings of the mulberry, *Morus multicaulis*, became a speculation that could be equalled only by the heyday of the Dutch tulip craze just two centuries earlier. At the height of the boom, mulberry cuttings only one-year old sold at from $2.00 to $5.00 each, and the demand could not be half supplied. A very few years later mulberry cuttings went begging at a penny apiece. Possibly in all history there are only two or three other cases where high hopes and boundless enthusiasm brought forth so little.

It must be that tanning leather is one of the most ancient arts. It is, in a way, something decidedly more than merely a handicraft. The maker of grain cradles must be a most accomplished worker in wood, but, after all, his task was one where his work was open and always beneath his hand. In it there was nothing of the occult or mysterious. But the tanner, though perhaps he only dimly comprehended it, was carrying on a traditional technique dealing with complex and little understood processes representing the accumulated experience of a long line of predecessors. Like the making of cheese or the baking of bread, his art was dependent upon certain chemical reactions which he could hardly be expected to understand or explain but which he had learned to control. Above almost any other vocation, his procedure was based upon time-honored rules of thumb.

Long before man devised any kind of woven fabrics he stripped off the skins of the animals and shaped them into coverings against the cold. He could hardly fail to note that rawhide had serious limitations; he experimented with treatments that might improve it. Eventually he hit upon certain processes which gave him a product better suited for his needs, and so it is that throughout history the tanner has had an important place in the contemporary economy. His was a time-honored craft. Nine centuries ago lived William the Conqueror, who was the result of his father's dalliance with a tanner's daughter. His lowly and irregular origin was never forgotten, and the story runs that once when he was besieging a French town, the defiant inhabitants hung hides over the walls with taunting cries "Work for Tanners." Today, tanning is one more process of the machine age. It is carried on in a relatively few large establishments equipped with astonishing mechanical devices; it brings to its aid a very high order of chemical science and technical research; and it goes to the corners of the earth to bring back an astonishing variety of products.

Tanning during the years of which I write was carried on in a great number of tiny establishments, each of which served the needs of a circumscribed community. The tanner himself possessed no book and his scientific training consisted of a body of traditional methods which were an inheritance from those who had gone before. Everything that he needed in his craft was obtainable at hand. His work seemed satisfactory and sufficient in his time. In those halcyon days of the rural handicrafts, the farmer was not only fed and clothed from within his own fence lines but he and his family went shod in footgear derived from his livestock. In early winter he killed a cow or heifer to furnish beef for the coming months and preserved it by "corning," drying, or smoking. The hide, along with sundry calf (kip) skins and sheep pelts which may have accumulated, was carried to the community tanner. From it the entire family was outfitted with boots and shoes, while the local harnessmaker, as occasion required, made it into harnesses for the farm team. Not infrequently the master himself bestrode a saddle built from hide of his own production, and on bitter winter days he went snug and warm in a jacket of sheepskin, with the unpulled wool turned next to his body.

From the early beginnings of America until the midyears of the nineteenth century, tanning was carried on in a multitude of small establishments. In 1845 there were 1,414 tanneries in our state; there was hardly a township but had one (or more) representative of this industry. The community tanner held his own better and longer than did the grower of flax or the local manufacturer of wool. In 1855 the state still counted 863 tanneries and, strange to say, these diminished in number only slightly during the next ten years for there remained 820 in 1865, and there was no rural county of the state but had several. It would appear that the Civil War brought about a temporary revival of flax and I think that it had a comparable influence on the rural tanneries in that it kept alive a dying industry. Even as late as 1873, my own rather small and very strictly rural county of Schoharie still listed thirteen tanneries in the local county directory. This, however, was the final stand of the currier, and I think the last of them has now been extinct for more than fifty years. Their passing is only one more symbol of the departure of the homespun age.

Most of these old-time tanneries were exceedingly primitive affairs, the whole labor force frequently being represented by one man who was both owner and general superintendent. Sometimes the proprietor was a farmer by vocation and a tanner on the side. In two local instances within my knowledge, a third calling, that of harnessmaker, was added. Such a multiplication of vocations was a perfectly proper and natural development of the handicraft age which decreed that a man must never be idle. Indeed, harnessmaking fitted rather well into the scheme of things because tanning itself was almost dormant during cold weather. Of course if a man could make a harness he could work leather in other ways, and my father had a wallet made by one of these farmer-tanner harnessmakers.

Rawhide might be tough, but it would stretch out of all reason when wet and it became miserably hard and horny when dry. Then too, it would decay very quickly if moist. It was, in general, wholly unsuited for boots or harness. Only when the gelatin of the rawhide had been united with the tannin of oak or hemlock bark did the product become suitable for the multitudinous uses to which leather is put. The mechanical equipment of the pioneer tanneries was exceedingly primitive. Doubtless, at one time the bark was simply broken by pounding it with heavy beetles or mallets, but even before the machine age there was in common use a bark mill which in its general construction was wonderfully like a greatly overgrown old-fashioned coffee mill. This bark mill was fitted with a horse sweep and so the bark was crushed. It was said that bark should be ground the size of wheat kernels, but this was an ideal seldom attained.

When I was a very small boy, Kilfoyl's Tannery was still making feeble efforts to survive in a world where it really had no place. It was only a short mile up the road and just beyond the far end of the farm, so it was not really foreign territory. One of the most shadowy recollections of my early years is the rather patriarchal figure of Thomas Kilfoyl as he stood by his mill throwing in pieces of bark while he urged on his patient and somewhat decrepit steed. Kilfoyl was a fine, white-bearded Irish gentleman, and a maker of honest leather of specially good repute, but he outlived his business and left no successor. As a boy, I had the habit of cultivating the acquaintance of old men. This ancient tanner was on my list of friends and he might have taught me so much. I wish that I might again hold converse with him of a summer's afternoon!

The old-time tanner's work was to a considerable degree seasonal. It was late fall or early winter when most of the farm butchering was done, and most of the tanner's raw material came to hand. Just as the miller almost universally tolled the grain which came to his mill, getting every tenth bushel in lieu of cash payment for grinding, so tanning was most commonly done on shares. The finished leather was equally divided between the farmer and the tanner. Thus the tanner from time to time accumulated a stock of finished leather which could be taken to some center of trade and sold for cash. Note, however, that no money had passed between him and the farmer, which again shows that pioneer economy was based largely on a wide system of share and barter, and the amount of cash in circulation was unbelievably small.

When the man who had just butchered a cow brought her hide for tanning, the very first procedure was for the tanner to take his razor sharp knife and on the flesh side of the hide, close to the point where the tail was joined, trace the initials of the owner. This made the equivalent of an indelible laundry-mark and no matter how many different processes the hide might go through, a year later as finished leather it could be handed back with the confident declaration: "This belongs to you." Having established this permanent identification mark, the tail, legs, and ragged head were trimmed off as not worth tanning. These fragments, however, were salable to the tin peddler, that once-universal scout of trade, now absolutely extinct in the countryside. Calfskins and sheep pelts were tanned whole, but cowhides were split straight along the back line making two identical "sides," the term which from now on would always be applied to them.

Aside from his one-horse bark mill, the old-time tanner had no mechanical helps except a few simple hand tools. His first step was eliminating the hair or wool. Sheep and lamb skins were moistened and stacked in piles so that they would sweat. This loosened the wool so that it could be pulled. I have already set down how a century ago, Betts Brothers made felt hats out of the wool thus salvaged from their tanning operations. I have heard that cowhides could also be sweated but certainly the more usual plan was to put them in a vat of milk of lime which loosened the hair so that it would slip. This cowhair, by the way, had a definite trade value, being always in demand by masons to mix with the plastering mortar made of lime from the neighboring limekiln and sand from the local sandbank.

Having removed their hair by use of lime, it was of prime importance in securing good, soft, durable leather that the last

The Tanner

vestige of lime be removed before proceeding any further, and to do this seems to have been no easy job. The hides, after being scraped free of hair, were removed to another vat where they were put to soak in a witch's brew, known as "bate," which was a mixture of hen dung, salt, and water. I do not pretend to explain the efficacy of this strange broth, but it was the traditional formula among old tanners and its universality attests its efficiency. After a period in this vat, the hides were very thoroughly scrubbed and rinsed in pure water, and it was widely held that the character of this water was important, soft water being preferable to hard.

Then the soft, plump hides were ready for their long immersion in the tanning liquor. There is a long list of vegetable tanning materials, to say nothing of various chemicals which will do somewhat the same work. Sheep skins were often tanned (or, more correctly, tawed) with the wool on, the procedure being to rub the inside surface with various mixtures, a combination of powdered alum, salt, and ashes being a commonly approved formula. For the tanning of cowhides and calfskins, the use of hemlock or oak bark was practically universal. Hemlock bark was much more readily available, but there was a belief that for harness leather oak was superior.

Even a small tannery needed several vats. They were each about six feet square and four feet deep, built of plank, and sunk into the ground so that their tops were only a little above the floor, with narrow passageways between. I have been told that life was one long nightmare for the mothers of tanners' children because of a haunting fear that they would fall into a vat and drown.

In the actual procedure of tanning, a layer of ground bark was spread over the bottom of the vat and covered with a hide. On top of this would be spread two or three shovelfuls of bark and then another hide, and so on until the vat was filled, when water was put in to cover the topmost skin. From time to time, the hides were examined, repiled, and if it seemed wise, more bark added. It is said that hides commonly lay six months in the tan vat; there was a general belief that prolonged soaking improved the product. I suppose it was this idea which gave birth to those delightfully whimsical lines from "The Deacon's Masterpiece" in which is described the leather that entered into the construction of that fabulous vehicle:

> Boot, top, dasher, from tough old hide
> Found in the pit when the tanner died.

In any case, it was often a year from the time the hide was brought to the tanner until the finished leather was ready for delivery.

Practically all the tanner's work was direct manual labor, almost unrelieved by mechanical aids and some of it, notably "beaming," that is, removing the surplus fat and connective tissue with the aid of the fleshing knife was particularly exhausting toil. After the tanning process was completed the leather must be thoroughly washed and then dried by hanging over poles in the loft. It was important that the drying process be not too rapid and yet with enough access to air to avoid molding. The need of space for drying made it necessary that even a small tannery should be quite a sizable building. When sufficiently dried, the leather was scraped, rubbed with a mixture of tallow and neat's-foot oil, blackened on the grain (hair) side with lampblack, and then rubbed smooth to give something of a polish. I am told that modern tanning has almost wholly substituted chemicals for hemlock bark and that warm tanning liquor and other short-cut methods have reduced the time required to a small fraction of the old period. Indeed, I have just read that it is now possible to convert a fresh cowhide into finished leather within eight days. Modern tanning is doubtless fast and efficient, yet when we come to consider real leather quality—softness, strength, and durability —there is at least a very general belief (in which I am glad to share) that the product of the old-time art was superior.

Two generations ago, it would have been easy to find graybeards who could have told in authentic detail of the procedure followed in the days when competent experts were tanning leather in the crossroads tanneries scattered everywhere over our farm country. Today, it is hardly possible to assemble any precise account of the primitive tanner's lore. Within my lifetime, definite memories have become only fading tradition. I must confess that what I have here set down represents only certain neighborhood customs which were familiar to my father along with the dim half-remembered accounts of Thomas Kilfoyl. In addition, there are certain bits from James Baldwin, an ancient man whom I have already mentioned more than once. He was not a tanner but in his time he had peeled a good deal of hemlock bark, and he was a Nestor who knew something about almost everything that had to do with the homespun age.

I have been unable to get much information as to the price of hides and leather during the first half of the nineteenth century. During the decade between 1840–1850, the *American Agriculturist* and *The Cultivator* in their farm market reports, quote "Hides —dry Southern" at a price usually varying between six and ten cents per pound, a very low price for a dry hide. It is certain that even at that date, tens of thousands of hides were imported from South America. This is surprising in view of the very large number of domestic skins available. It certainly indicates that leather found a very large place in the pre-Civil War economy.

In the Report of the American Institute for 1847 is included a very interesting discussion of tanning by Z. (Zadoc) Pratt of Prattsville, Greene County, N.Y. Pratt was at that time the proprietor of a tannery said to be the largest in this country. His contribution proves him to have been a writer of clear and vigorous English, as well as a businessman who kept a minute account of all his operations through many years. In addition to this, he was a genuine student of the technical problems involved and he possessed the fine trait of willingness to pass on his knowledge to others. Apparently, he had no trade secrets to conceal. From this report, which I am sure represents the most approved practices and experiences of his time, I transcribe a few outstanding observations:

During the years 1827 to 1847, the average price of sole leather in New York City ranged between nineteen and one-half cents in 1831 to eleven and one-quarter cents in 1846. Pratt's establishment tanned more than 60,000 "sides" per year—a tremendous business for that day. He had cut more than ten square miles of hemlock forest and found the average yield of bark was eighteen cords per acre. He estimates the cost of bark as only $3.50 per cord, but I learned from other sources that in those same years bark in northern Schoharie County sold for as much as $10.00 per cord, and this at a time when ten dollars was a large sum of money. His books showed that, on the average, he secured 194 pounds of finished leather (perhaps twelve to fifteen hides) from each cord of bark. He also records with much satisfaction that he had a bark mill which would grind a cord of bark per hour and that he operated it continuously both day and night. At the close of his discussion he says (and I cannot but admire him for this final touch) that all his far-flung activities extending over twenty years, were carried on "without the use of ardent spirits and without ever having a single side of leather stolen."

Zadoc Pratt was a great captain of industry in his time. In Pratt Park, high above the village where his life's work was wrought, you may today see, among many other curious mementos, his bust cut of a solid crag of the mountains. From below, the murmur of the creek comes up faintly, the surrounding Catskills heap themselves in wooded folds and the face of this old-time man of affairs and master of men gazes out forever with unseeing eyes over the scenes of his labors and his triumphs.

CHAPTER XVIII

Workers in Leather

INASMUCH as the old-time shoemaker was so directly dependent upon the community tanner, it would seem that now is the logical place to consider him and his contributions to the economy of his time. The neighborhood tanner and the rural shoemaker existed side by side from the early beginnings of this country and they made their exit from the stage at about the same period. As late as 1860, both might still be found in numbers carrying on very much as always, but by 1890 both crafts were practically extinct. The introduction of a sewing machine which could successfully stitch leather and could be applied to the rather difficult job of boot-and-shoemaking, was the beginning of the end of the worker who sat on his low leather seat with his stitching horse conveniently near and his wax end and assortment of small tools directly at hand. It was in the 1850–1860's that the sewing machine invaded the field of boot-and-shoemaking. At first the machine manufacture of footwear was centered in eastern Massachusetts, with Lynn as the capital of the industry. It is said that in 1860, more than one-third of all the factory-made shoes came from three Massachusetts counties.

There was a period when common practice of the business was to make the uppers and cut out the soles in factories and then distribute these half-finished products to fireside workers, who pegged on the soles according to the long-established practice of the craft. Thus there grew up an important countryside industry ministering to and dependent upon the large establishments which did part of the work. I remember hearing the late Herbert W. Collingwood, the lovable editor of the *Rural New Yorker*, reminisce concerning his boyhood memories of this. Then the day came when a man devised an incredible mechanical wonder which would peg more boots than twenty men. Collingwood said, "I remember the distress and dismay when the word went around that there would be no more boots to peg." Suddenly a community which had come to depend upon this particular activity found its livelihood vanished, save in the case of one man. The nub of the story was that the man who had made the wonderful machine that put so many folks out of business journeyed to this sole remaining shoemaker to have his feet measured for boots because the hand worker not only made boots, but made boots that would *fit*. This was Collingwood's little story, and of course it carried the moral that if a man could do one thing supremely well he would never be without a market for his skill.

In passing, let me say that for the sake of accuracy I propose to write boot-and-shoemaker rather than the easier term "cobbler," because by lexicon definition a cobbler is the repairman rather than the original maker. One might even suggest the cobbler be called a "shoe-tinker." It is true that we still have cobblers of a sort. Almost any good-sized village or city may have a shoe shop, called perhaps the Electric Shoe Repair Parlors. These places will have at least a shoe sewing machine, and will mend a rip or put on a "live rubber heel" or a neolin half sole while you wait. The modern representatives are an almost immeasurable remove from the primitive craftsman of three generations ago who, out of home-tanned leather, hand-twisted flax, wax-end, and wooden pegs, and over a last which he himself had made, shaped the boots and shoes for his patrons. I am sure that of all our folk crafts there is none more ancient, more necessary, or more universal than the trade of which St. Crispin is the patron saint. Its votaries once flourished not only in every township, but in almost every school district.

Roving Cobbler

That oft-quoted census of 1845 counted the establishments in several lines of rural industry, but for some reason the shoe shops were not enumerated. The explanation may be that at this date shoemaking was regarded as wholly a one-man craft rather than a manufacturing industry. Ten years later, however, there were reported 1,463 "boot and shoe shops." A decade later this number had fallen to 525. Evidently the industry was consolidating into fewer and larger establishments. In the opinion of the census taker, the man who sat on his bench in his own living room did not constitute a manufacturing establishment. Concerning the number of these individual workers, pretty conclusive evidence can be gathered from the various county directories and gazetteers. Ordinarily, the county directory was a publishing venture based on subscriptions for the forthcoming volume. The usual price was one dollar or a dollar-and-a-half and this included a line giving the subscriber's name and business. In a word, it was a local "Who's Who" confined to those who paid for the publicity. In Child's directory of Schoharie County for 1872–1873, it appears that there were no less than ninety-six men who were willing to pay to be listed as "custom" shoemakers. By the same token, we had 131 blacksmiths. That was more than three-quarters of a century ago. Since then, the very last of the cobblers has laid aside his tools forever and left no successor, while of that goodly company of blacksmiths there are now less than half a dozen who make any pretense to represent the fading traditions of the craft.

The use of wooden pegs to fasten the taps to the uppers, while at one time almost universal for the footwear of both men and women, was by no means the earliest method. Indeed, it is said that wooden pegs were first used in New England about 1812 or, according to another account, 1818. At any rate, their use soon became exceedingly common. At first the bootmaker whittled (or split) his own, but shoe peg factories were soon established. In 1841 a factory at Laconia, N.H. is said to have produced nearly all the pegs used in this country. They were made from black, yellow, or white birch and from hard maple. The daily production of pegs exceeded fifty bushels. This establishment did not, however, have the field to itself for long; in 1850 the census returns for Massachusetts showed a total production of 17,800 bushels of pegs worth $12,900. This seems very little money for such a tremendous output, but it was the day of cheap labor and abundant raw material. In 1855, New York State reported fifteen shoe peg factories, but ten years later only three survived. Cobblers bought their pegs by the quart in several different sizes and the price for that unit of measurement was only five or six cents.

The ankle length shoe (for men at least) is a fairly recent innovation. The bootmaker of the homespun age knew little about low shoes. In his time practically all men wore boots reaching to the knees. The farmer wore heavy cowhide for every day, but he had a pair of fine calfskin for church and social occasions. The idea of a shoe open in front and laced with a shoestring is said to have originated with a Yankee cobbler about 1791. It seems certain that well into the midyears of the nineteenth century, the practice of country shoemakers was not to make footgear "rights" and "lefts" but to make the pair over the same last so that both shoes were identical. If the owner would conscientiously transpose his boots from one foot to the other each time he put them on, they would never run over at the heels or wear out the taps unevenly. My father remembered and frequently told me of this careful economy of his boyhood. I think it safe to say that up until the close of the Civil War the average man and woman of the farm country went shod in the product of the local cobbler's art, and even after factory shoes were generally introduced there were many conservative men who steadfastly refused to wear "store boots."

There is no question but that when the homespun age was in flower, literally thousands of shoemakers sat on their low benches and industriously tap-tapped-tapped or drew out the long waxed ends from morning to night that our farm people might be shod. The last man who ever practiced his craft in my community was one Fred Martin. It must be about seventy years since he ceased, so that while I can say confidently I remember him and his shop my recollections are blurred and indistinct. The house where he lived still stands, and today's living room, which was his shop, has an unusually large window built thus so that

Fred's bench would be flooded with light. I think it proper that I might here speak of Fred Martin's unfortunate venture into agriculture. Through long years of patient industry he had accumulated what to him seemed a little competency. How much, I do not know, perhaps two or three thousand dollars. Then, in his folly, Fred must needs leave his craft, which he understood, and buy a farm, about which he knew nothing. He purchased a poor hill farm just as land values began their long decline. In his new occupation he was helpless, his equity was soon wiped out, and he was forced off the farm, a sadder and a wiser but unfortunately an aged and broken man. This back-to-the-land foolishness is responsible for too many economic tragedies such as this.

The workshop of the old-time cobbler was very frequently a corner of the living room (he required little floor space), but there were also itinerant workmen who went from farm to farm with their bench and kit and wrought beside the fire of their employer. If the family was large, it might be several weeks before all the members of a household were properly outfitted. In the days when there was very little contact with the outside world, the roving cobbler, especially if he were skilled in pleasant gossip, was welcomed both as an employee and as a guest. Then, too, inasmuch as his clients were all conveniently on hand, it was easy to cut and fit and try as the work proceeded. My father lived long enough ago to remember how the local tanner would be scolded and importuned to hurry and get the leather finished because the shoemaker was expected.

In the log house of the D. P. Witter Museum on the State Fair Grounds at Syracuse in 1926, one might have seen Mr. John Mulberry of Potter's Hollow actually wearing a pair of boots made after the exact methods of long ago. He is my authority for the statement that at an early date the cobbler shaped his own lasts. Sometimes when he established himself for a siege of family cobbling, he measured the biggest foot of the family and then made a last to match it. This having served its purpose, the shoemaker proceeded to shave it down and shape it to fit the next largest foot, and so on until every member of the family, including the six-year-old boy, would have his footwear built over the self-same last, by this time reduced to a fraction of its original size. This is the tale as Johnny unfolded it. I confess that to me it seems a bit apocryphal.

Johnny Mulberry said, too, that as late as the 1870's the coarse boots of the outdoor worker usually cost about $3.00 per pair, but much less if the leather was furnished by his patron. A pair of fine kip or calfskin boots cost $5.00 or more, and on these the maker lavished special care. I have seen a pair of very old boots which had a lining of leather sewn to the outside by stitches that took hold of the outside leather but never pricked entirely through or showed any indication of the seam. To do a job like this surely called for patience and skill of a high order. Mulberry told me that in his prime he could make out of the raw three boots (a pair and a half) a day, but it may be well to keep in mind that old men remember the maximum rather than the average day's work.

The shoemaker at his work sat on one end of his very low bench, which was only about eighteen inches high. The regulation seat was a circle of leather nailed over a round hole. The shoe-maker hunched over his work, some of the time with the half-completed boot held against his chest. For this reason his calling was commonly believed to be especially unhealthy, and I find that about 1860 one writer urges this as a reason for the introduction of factory methods. As a matter of fact, I believe that cobblers, beyond other men, were inclined to grow stooped and round-shouldered with the passing years. But perhaps this was not as fatal as we might expect. A generation ago died one David Cronk whose home was in Lewis County. Not only had he attained the ripe age of 105, but, in addition, he had the honor of being the last survivor of the War of 1812. Most of his long life he had spent on a cobbler's bench, but it will hardly be asserted that as the result of his unhygienic occupation he was prematurely cut off in the flower of his youth.

The antique hounds have long been spoiling the countryside of spinning wheels. More recently they have added the collecting

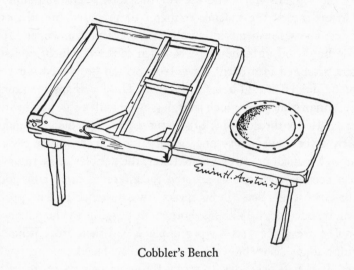

Cobbler's Bench

of cobblers' benches. In the heyday of the cobbler's craft there must have been literally thousands of them, and I judge there still remain a great many stored away in sheds and attics all over eastern America. The bench was perhaps five feet long; at one end was the round, sunken leather seat, while the other end was divided into a large number of compartments to separate the various sizes of pegs and nails. The cobbler's tools were simple and inexpensive, but a full kit had a surprisingly large assortment. There were many sizes and shapes of awls, as well as three or four thin knives, one or two of them curved, and all kept as keen as a razor. For this purpose there was always at hand a fine whetstone and a strop. The cobbler's wax ends and his ball of shoemaker's wax were within easy reach, while at his knee stood the stitching horse whose jaws clamped and held the leather in position for sewing. This left both hands free to punch the holes with an awl and then draw through the threads, a stitch entirely unlike that employed by a woman sewing cloth. Among his tools was one small gadget to which he gave the fanciful name of "the nigger's lip." The use for this tool is now unknown.

In order that the shoemaker might be prepared for all sorts of customers, for men with number ten feet, for women, and for little boys and girls, he had a wooden rack on which his lasts were arranged in orderly rows, from which he picked out the one he judged would best fit his client. Until perhaps a century ago, and in some cases much later, he sewed with a stiff hog bristle waxed to his thread. This bristle served as a needle, and quite likely a very satisfactory one, because it could readily be passed through a small or curved hole. Later, the cobbler came to have steel needles with curves adapted to different parts of his job. The tap (never "sole" in his speech) was fastened to the uppers with wooden pegs. With his short, stout pegging awl he punched a hole for each peg, set the peg in place, and then drove it home with a single smart blow of his light, broad-faced pegging hammer. If it was necessary to strike the same peg twice, by that token he was emphatically a botch and no cobbler.

Against his wall hung tin or wooden patterns, which he laid on the leather and then with his keen knife cut out the odd shaped pieces of leather, which when crimped into the proper relationship became the uppers of a boot or shoe. In the corner were flung pieces of leather along with rolled-up sides which farmers had brought to be shaped into footwear for their needs, while over all hung the not-to-be-forgotten aroma of tanner's oil and hemlock bark.

I think it proper that I should here set down a word of tribute to Johnny Mulberry, whom I came to know and greatly esteem. When I was associated with him he was a timeworn, withered man whose life had been lived in simple, humble, perhaps in narrow, ways, but there have been few men more richly endowed with the graces of contented and kindly judgments and warm-hearted friendliness. He must have been one of the very last of the old-time breed of cobblers and it was his fate, along with his contemporaries, to be caught in the cataclysmic upheaval of the new order. I like to say that with his work on the New York State Fair Grounds in the 1920's a very ancient craft came at length to an end.

There is one other craft that had a very intimate connection with and dependence upon the tanner's calling and is, therefore, most logically considered just at this time, viz., the calling of harnessmaker. This was an art once quite indispensable to the rural community although it must be that in the number of followers or the value of its product, it held quite a subordinate place as compared with boot-and-shoe-making. In a word, providing footgear for the countryside was bigger business than making harness for the farm team.

Of course these two vocations were wonderfully closely related. Both of them were fundamentally concerned with leather, and keen knives, waxed ends, and stitching horse were the almost identical tools of each. I have already said that my father knew two men within our locale of three or four miles who were tanners by vocation and harnessmakers incidentally, or possibly it was the other way around. In any case, agreeable to the usual custom of tanning on shares, the tanner always found himself with leather available and if he could convert it into harness, it was doubtless a better business venture than selling in Albany, which was our market for such commodities as could not be bartered at home. I never happened, however, to hear of any one who combined the dignities of shoe-and-harnessmaker although the two enterprises might very easily and naturally merge. In one respect the two skills were aiming at quite different ideals. The shoemaker, presumably at least, was making a shoe to fit a particular foot while the harnessmaker's model was exceedingly elastic because he planned a contrivance that could either be let out to fit the biggest horse on the farm or be taken up to suit a much smaller animal.

The workaday procedure of the two craftsmen was somewhat at variance. Nearly all of his working hours the shoemaker sat on his low bench. At first sight, his seat seemed too low for comfort, but apparently long generations of experience had demonstrated that this particular arrangement best suited his job. The harnessmaker spent a large portion of his time standing by his broad, flat-topped table on which a side of leather could be spread, while he with rule and straightedge cut straps of the proper length and width with nice judgment as to what portion of the hide worked best for the different purposes. He had to be familiar with the intricacies of trace, lines, backstraps, bellybands, "britchen" (breeching), crupper, headstall, back pad, holdback straps, and surcingle, but it might seem that he hardly needed the accuracy of the man who must shape leather so that it finally emerged to fit the human foot. In a way shoemaking still remains an occupation, even if the cobbler has become just one more worker in a big factory where his function is to tend a machine. But the harnessmaker, along with the village blacksmith, seems to be fading into oblivion. The horse on which their existence depended is rapidly becoming a zoological curiosity.

It would seem that we shall never be able to assemble any accurate information as to the number of craftsmen who carried on when the homespun age was at its zenith. That invaluable census of 1845 gives the number of farms along with a really astonishing mass of statistics relative to the acreage and total production. So far as agriculture is concerned, we have a pretty good picture of the times. Concerning industry, however, the figures for that year are much less satisfactory. The report includes some very rudimentary statistics relative to just twenty-one industries and ignores all others. Those chosen include sawmills, gristmills, tanneries, breweries, distilleries and asheries, but not shoe factories. Apparently at this date the individual cobbler still held the field, a conclusion that would be definitely in line with my father's memories. These twenty-one industries are given by location and also the dollar value of their production, but not a word as to the number of workers employed.

Ten years later, in 1855, the State of New York made what for that era must have been a most complete occupational census and specifically enumerates no less than 374 different vocations, callings, and professions. However, this most exhaustive over-all inquiry leaves the most interesting inquiry still unanswered. To what extent had these counted workmen found employment under the banner of the oncoming industrialism and to what extent were they self-employed freemen still carrying on as in the past? My own belief, which is nothing more than an impression, is that until the outbreak of the Civil War our thinking and planning were in terms of the old regime.

CHAPTER XIX

Some Minor Crafts

THERE are certain minor crafts of the pioneer era which deserve at least brief mention in passing. Among them the most important and best remembered, are pottery, brickmaking, lime burning and basketry. These arts have been a part of the farm picture almost from the beginning. None of them may be described as dying, and all of them may be expected to continue indefinitely into the future, but in a manner very far removed from the methods and customs of the pre-machine era.

Concerning pottery only casual reference need be made. Old-time earthenware, like old glass, early clocks, or antique furniture, has been taken over by the collectors and become a field of detailed special knowledge. Anyone who desires to pursue these subjects will find no lack of books. It is commonly said that the first potter in America was one Daniel Cox, who wrought at Burlington, N.J., as early as 1684. It would appear that at the date of our first occupational census (1855), the potter was a relatively unimportant craftsman in New York State. Only 287 persons reported this as their occupation. The primitive industry, however, was widely scattered. Potters were found in some thirty-six counties, although at least eight of these counties reported only a single representative of the craft. In these cases it was very evidently a one-man enterprise where some lone worker on the potter's wheel, out of native clays, was making the bowls, basins, pancake "crooks," and baked bean pots for the local countryside. It is a far cry to the potteries of today.

The potter's wheel, one of the most ancient of mechanical appliances, is almost extinct in commercial work. Modern earthenware is molded rather than "thrown" on the wheel. Even today, however, we now and then come on a man who can still demonstrate this most fascinating of manual skills. Beholders have always been intrigued by the fashion in which the clay was made to obey the potter's will. Jeremiah, the prophet, lived twenty-six hundred years ago, and he wrote "O house of Israel, cannot I do with you as this potter? saith the Lord. Behold, as the clay is in the potter's hand, so are ye in mine hand . . ." (Jeremiah 18:6). Omar Khayyam flourished nine centuries ago, and he, too, has paid tribute to the potter's art:

> It chanced into a potter's shop I strayed
> He turned his wheel and deftly plied his trade
> And out of monarchs' heads and beggars' feet
> Fair heads and handles for his pitchers made.[1]

The settler on the fringe line of the white man's occupancy built him a log house. His Iroquois forerunner seems to have been content with a multiple dwelling bark house with a fireplace in the center and a hole in the roof above it which, when the wind was right, took care of part of the smoke. The paleface newcomer insisted on a fireplace with some sort of chimney. In the very beginning, a stick-and-mud flue was made to serve. When he laid up a chimney of rough fieldstone, it symbolized advanced and

[1] Quatrain CCLVI of the Whinfield version (New Century Library; New York: Thomas Nelson and Sons, n.d.).

permanent construction. Soon, however, often within a dozen years, the log house would be replaced by a frame dwelling. Then there arose a need for a brick fireplace and chimney, not to mention the Dutch oven also built of brick. Then, too, a good many years ago it was not uncommon, both in the villages and among prosperous farmers, for men to declare their prosperity and social standing by building substantial homes of brick. Throughout the more fertile regions of eastern America are brick houses which in many instances were built within a generation or two after the first settlement. I have already written of how my own great-grandfather was born in a log house but when he was twenty-one the family moved into a commodious and dignified home built of brick molded out of the clay of the home acres.

So it was that just as soon as the new country was permanently occupied, the ancient and primitive art of brickmaking flourished in a multitude of places. As a case in point, I will cite the fact that on my farm there is one point in a level field where the plow will never fail to turn up red soil and abundant fragments of brick. Without any question, this little area is an abandoned brickyard. The rather astonishing circumstance is that, while my father's memory surely embraced the farm happenings and conditions of the 1840's, he had absolutely no knowledge concerning when brick was burned here or the extent of the operations. Here was an industry that arose and perhaps prospered, and which surely passed away leaving behind not even a tradition. Perhaps in my boyhood, if I had inquired, I might have found some old man who could have unfolded the detailed story. Now it can never be known. A long half-mile in the opposite direction from our farm brickyard there is a still distinctly visible depression marking the spot whence was taken the clay to make the brick for the Dutch Reformed meeting house in this hamlet of Lawyersville. This was the "new" church built in 1850. Viewed today, the size of the excavation would suggest that operations must have been considerably more extensive than the erection of a single building.

Just in passing, I am going to express a large measure of skepticism relative to the many old houses allegedly built of brick brought over from England or Holland. There are variations of this tale, but almost any pre-Revolutionary brick house is apt to be glibly accredited as built of brick brought across the ocean. Once put in circulation, such a statement goes happily down the years; no one has either the knowledge or the wish to make specific denial. Now I cannot explicitly declare that there are no "overseas" brick in our old coastal cities or even in our Hudson River towns, which in all essential respects have always been seaports. But when the claim is made in behalf of houses perhaps fifty miles from tidewater, it becomes fantastic. The transportation difficulties of that day were bad enough; no man could have been quite so bereft of intelligence as to transport brick for fifty miles over dirt roads with horses or oxen for motive power. On the other hand, serviceable brick could be burned from almost any type of clay and clay beds were everywhere. The equipment required was hardly anything beyond a few wooden molds such as could be turned out by any country carpenter. Wood for fuel was easily obtainable, and unquestionably the countryside was not devoid of men familiar with the simple technique of the brickmaker's craft. Undoubtedly among the first colonists there were men who had been brickmakers in Europe. At the time of our first occupational census there were in New York State 1,627 men who declared that by vocation they were brickmakers. It should be remembered that by this date machinery (pugmills for puddling the clay and brick molding machines) was coming in, but a generation earlier the craft was purely hand labor unrelieved by any mechanical aids.

I have heretofore referred to my great-grandfather and the brick house he built on the farm where, as a young man, he had located when it was still untouched wilderness. In the Farmers' Museum at Cooperstown may be seen one of the molds used to shape the clay. When the house was finished, this mold was laid away in the attic, there to remain untouched for 120 years. The mold is nothing more than a shallow wooden box, two and one-fourth inches deep, about a foot wide, and two and one-half feet long. One partition lengthwise and two partitions crosswise divided it into six compartments, each the exact size of a brick. In use, this mold must first be thoroughly wet, then sanded by sprinkling sand on

"My Great-Grandfather's Good Brick House"

all surfaces touched by the clay, and then tamped a little more than full of properly tempered clay.

The surplus clay was then cut off with a straightedge and the box turned upside down on a board. Then a little tapping would cause the six green brick to drop out. As soon as they were dry enough to endure careful handling, they were put on trays and carried to be piled on racks where they had free access to air but were protected from the sun and rain. This curing process continued until the bricks were well dried out. When ready for firing they were piled to make a kiln, which was a square pile laid with flues near the ground which could be kept stoked with cord wood. The bricks had to be laid up, not tight, but with a little space between so that the hot gas from the fire could circulate through the mass. As compared with modern methods this seems an almost incredibly crude process. It commonly took a week to burn a kiln of brick and the ideal was that all parts of the pile should be brought to a cherry red heat, something that was rarely achieved in practice. When the kiln had cooled and was taken down, the soft or "slack burned" brick were sorted out and added to the next kiln for further firing.

There is also another mold which made only two bricks, or, more correctly tile, at a time. Each brick was eight inches square and somewhat thicker than the regular kind, and these were used to build the fireplaces and hearths. A casual inspection of the old house leads me to believe that these long-dead brickmakers did a good job. I have always understood that John McNeill's four sturdy sons had a very large part in the work. This was almost a century and a quarter ago and in a remote New York countryside, but the dimensions of the brick were almost exactly the same as those made today by machine methods. Is it not passing strange that in a world whose most unchanging characteristic seems to be endless change, the dimensions of the humble building brick should remain immutable?

There is one specialized rural activity which should be mentioned and most appropriately just at this point because it is so closely akin to both pottery and brickmaking. I refer to molding and burning tile for agricultural drainage. Possibly this ought not to be considered an activity of the homespun age because it did not become a widely adopted farm practice until well in the 1850's. Like many other important advances, its beginnings seem very largely to center around one crusading man. John Johnson was a Scotsman who came to Geneva, N.Y. in 1821, and purchased a farm of 112 acres. He was a man of vast agricultural enthusiasm and it would seem that he must have possessed at least his fair share of traditional Scotch thrift. His farm, while fertile, lacked good natural drainage and, remembering the use of tile in his old homeland, he ordered a shipment of Scotch tile which reached New York in 1835 and from there was shipped to Geneva via the Erie Canal route. These were the first tile ever laid in this country, and Johnson is sometimes hailed as "the founder of tile drainage in America." Eventually he underdrained his entire farm with lines of tile laid in ditches two and one-half feet deep and only 20 feet apart. The results surpassed his hopes and on three different occasions in articles published in *The Country Gentleman* he wrote that he had made his land "too rich for wheat" because it grew so rank that it lodged before heading. The drainage practices he introduced were widely copied by the best farmers and before many years a number of establishments were making drain tile out of local clay for the use of the surrounding countryside.

John Johnson lingered on the scene until 1880 and was for many years a fine, outstanding figure of a man and a sort of elder statesman in New York agriculture. Few men so well deserve a place in the farm hall of fame. Indeed, I sometimes wonder if we have among us today any plain men of the soil who possess the enthusiasm, high hopes, eagerness of outlook, and the open mind which characterized certain farmers before the present century came down the stream of time.

Before stone houses could be built or brick chimneys laid or farmhouse walls smoothly plastered, it was necessary that lime be burned, because lime in combination with sand has through all time furnished the building mortar for the world. Even the coat of whitewash that was the first interior adornment of the log house was impossible until the housewife could get a few lumps of quicklime. It is a safe assumption that lime burning was an industry which flourished just as soon as the frontier settlements were made.

Unfortunately, there are some rather extensive regions that have no limestone. In some cases near the sea coast, the colonists burned lime out of clam and oyster shells. In other localities, far from limestone outcrop, it was necessary to team this indispensable product from what must have seemed far-off sources. In those localities where limestone was readily quarried, kilns were exceedingly common. Time was when we had at least a half-dozen within half as many miles, but the last of them has now been abandoned and forgotten for more than fifty years.

The old-time kiln was usually dug out of the side of a steep bank so that it was easy for a cartload of stone to be drawn to a point level with the top. Doubtless, the dimensions of the kiln varied with the ambitions of the builder, but a diameter of eight or ten feet and the same in height, with a flue opening at the bottom, may be considered typical. The kiln wall might be laid up of the prevalent limestone, but if so there must be a lining of fire resistant sandstone.

The operation of such a kiln was very simple. To begin with, an arched flue of limestone was laid at the bottom and then the kiln filled with chunks of limestone of the size experience had shown was best. The favorite fuel in the old days was dry sawmill slabs, which were fed into the bottom flue, and the flame and heat passed up through the overlying stone and out at the top. Commonly, such a small kiln would be fired for three days at a temperature considerably less than that required to burn good brick. It was two or three days before the almost red-hot mass cooled enough so that it could be comfortably handled. Lime was sold by the bushel rather than by weight, and the price ordinarily ranged between twenty and thirty cents.

In 1855, there were in New York State 129 men who declared that their vocation was lime burner. It seems certain that this relatively small number by no means indicates the importance of the industry. Lime burning was, like shingle shaving and charcoal burning, an important industry and yet, few men considered it their main trade. There were in our state at that period 1,031 men who preferred to call themselves quarrymen, and very many of these were familiar with making lime. The two operations were intimately associated and overlapping, and there is no doubt that many men classifying themselves as quarrymen would have been competent as lime burners. Such products as Portland cement, plaster board, and mortar shipped in a bag have made lime much less important than once it was. Lime burning has been taken over by a few big kilns operating continuously like blast furnaces and burning coal. The little community enterprise burning wood for fuel and making a few hundred bushels of lime as it was needed has become extinct. This is just one more example of the disappearance of a varied industrial life which was once a definite part of the rural community.

Pottery and brickmaking are surely related arts or, perhaps more logically, somewhat diverse developments of the same art. To these industries, lime burning may claim a certain kinship. All three require a very hot fire and plenty of fuel. Basketry, however, should be reckoned as part of the wooden age and closely related to the weaver's art. Basketry employs essentially the principles of warp and woof and it has been suggested that the first cloth appeared when some thoughtful savage asked himself why fibres could not be arranged in the same fashion that he had long used in making articles of wooden splints or twigs. Basketry, as a development among exceedingly primitive peoples, has been practiced over a vast period of time and with a great variety of materials.

The art had an important place in our pioneer economy. In 1855 there were in New York State 783 persons who reported their occupation was basket making. These were the folks who considered it their chief business. It is certain that in addition there were many others who could make baskets and did so as a sideline to their regular vocations. Baskets of diverse shapes and sizes had an important place in farm and household equipment until a period within the easy memory of elderly folk. In the old Northeast at least, the old-time worker wrought mainly with two different materials. For the home he made baskets of the peeled shoots of the osier willow and for the field and stable he made baskets woven from splints of white oak or ash. The willow basket is an amazing example of light weight coupled with astonishing strength. Nothing else made of wood can equal it in these regards.

The white oak or ash splints used in weaving baskets were made from sticks rived from a suitable log. The size of these sticks might vary through considerable limits, but approximately two inches square was usual. These sticks had to be thoroughly soaked in water and then vigorously pounded throughout their entire length with a heavy hammer until the laminations of the wood were loosened one from another so they might be separated into thin splints. It was a job calling for suitable timber, patience, a strong arm, and no small degree of judgment and skill. In passing, it may be mentioned that when Civil War years brought the first hay press ever seen in this locality, the bales of hay were bound, not with wire as has been the custom for many years, nor with the stout twine which the modern "pickup" baler uses, but with the same wooden bands as those used by the basketmaker. These were put around the bale like hoops and the ends locked together with a splice similar to that used by the cooper in making his wooden hoops. The labor involved must have been appalling, but it was an age when men did not rebel at toil.

Willow basketry still survives in clothes hampers, wicker furniture, and baby carriages, but the old-time basket, woven of oak splints pounded from the log, may be regarded as practically extinct. Nowadays a serviceable farm basket usually means a container of galvanized sheet iron. It is true, however, that never before have we had so many baskets as today. They are turned out in big factories literally by the millions as shipping containers for fruit and vegetables. These are flimsy contraptions, costing relatively very little money, and designed for only a single trip. If a large log of suitable wood be subjected to prolonged steaming it will become softened, after which it can be clamped in an overgrown turning lathe. As the log revolves a keen knife will peel off a thin endless sheet of wood, the so-called basket stock veneer. From this is made what we call baskets. Strictly speaking they are round crates, not being woven but, rather, nailed together. All this calls for a considerable investment in plant and equipment, and indubitably such an establishment is a development of the machine age. The old-time basketmaker who turned out oak splint baskets in various shapes and sizes as his customers might require is a figure now only dimly remembered by aging men.

There were two once widely distributed industries whose disappearance from our countryside I chronicle with much interest but without regret. In 1845 we had in New York State 102 breweries and 221 distilleries. Breweries were relatively unimportant. There were less than half as many and the reported value of their product was less than one-third that of the distilleries. Moreover, the breweries were largely concentrated in the sizable cities. Distilleries were a different story. Out of our sixty counties, forty-five had at least one distillery and many rural counties had several. The business was especially concentrated in Dutchess, Orange, and Ulster. The wide distribution of these distilleries and their usually small size indicates that to a great extent they were local utilities ministering to the local demand.

The 1845 census does not give the number of employees or the gallons manufactured, but it does give the value of their product, which is reported as approximately four and a quarter million dollars. Fortunately, we have at hand almost indisputable evidence as to the price of liquor at this period. The newspapers and agricultural journals carried a brief market report in which the price of whiskey was quoted just as casually as that of wheat, salt cod fish, wool, or dressed flax fiber. Apparently at this date the price was well standardized within a narrow range. For what is listed as "American Whiskey," the standard quotations ranged from twenty-two to twenty-five cents per gallon. A simple mathematical calculation involving the reported value of the product, together with the price per gallon, indicates that in 1845 this state manufactured some seventeen million gallons, more or less, of what was doubtless a very raw and potent beverage. The state at that date had about 2.6 million inhabitants and if this flood of liquor was all to be consumed within our borders, it represented an appalling per capita consumption. Perhaps much of it went beyond state lines, although at the same period some New England coastal towns were busily engaged in distilling rum from West India molasses, and the state of Pennsylvania was becoming renowned for its "hard licker."

I fear that if the truth must be told, the years before 1830 constituted a rum-ridden age. I mention this date because it marks the decade during which there was developed a somewhat active total abstinence movement. Life was exceedingly unsophisticated and surely the cocktail hour had not as yet been devised, but almost every community large enough to be deemed a village had its tavern, and many crossroad stores sold liquor as they did molasses. It was my father's proud boast that this hamlet where we live was unique among surrounding communities in that it never in its history had a barroom, although there was, as I shall particularize, a nearby distillery.

Perhaps the use of alcohol had a standing and respectability greater in the homespun age than now. Liquor was served (but to men only) at funerals and weddings, and on festival occasions, to say nothing of bees and barn-raisings. Even the preacher, making his pastoral calls, was not always scandalized if urged to take a "little something" for his stomach's sake. Sometimes this freedom of use eventuated in something of an orgy. At a Fourth of July celebration held at Sharon Hill in the early years of the 1800's, a Negro, one Mungo, was headed up in a barrel and rolled down hill, an experience he survived to remember if not to boast of for many years. While there were even then, some rigid total abstainers, most adult males drank, at least on occasion. At the same time, illogically, the use of alcohol was deemed exclusively a masculine prerogative, and the tippling woman was regarded as far outside the pale of respectability. Note again that at this period the manufacture of liquor was unhampered by any taxation or excise regulations and consequently it was available at exceedingly low prices.

I have just said that here in Lawyersville we had our distillery. It passed even before my father's memory, but I do know that it was in operation in 1819 and the stiller was one Thomas Wing. Also I know, at least within a few rods, the spot where it stood. It was not on the highway, but in a narrow hidden valley. I am sure that this was not for any purpose of concealment but solely because here was an abundant spring which would furnish water, not only for the mash tubs, but more important perhaps, to cool the condensing "worms."

Some few years ago I paid a visit to the site to see if there might be visible remains. I could find no foundation or cellar hole or any evidence that the spring had once been piped. Knowing the plumbing arrangements of the time, I have little doubt that a few lengths of pump logs conducted the water to the point where it was needed. The fact that there remains no physical trace is evidence as to how primitive and temporary the plant must have been. Nonetheless, there is conclusive evidence that once on a time it flourished. Every old farm needs certain names to designate the different fields. On this farm, in everyday speech, we speak of the twenty-acre lot, the spring lot, the pine grove lot, and the long meadow. The distillery was on a farm which joins our own and which had been held in one family line since the beginning of settlement. To this day the field running down into the narrow valley where the stiller plied his trade is in farm speech called the "still lot," thus embalming the memory of an enterprise which passed away a good deal more than a century ago.

There is a definitely folk tale relative to this establishment which it would seem might be taken at its face value. I heard it recently from the elderly great-grandson of the man who owned the farm when the still was a going concern. The farm has passed from that day to this by unbroken inheritance, so it may be said that the tradition has had the advantage of being always in family custody.

On this occasion a tub of mash was judged to have reached exactly the proper degree of maturity for distillation when the expert in charge made the somewhat disconcerting discovery that a prowling cat had tumbled in and found an alcoholic grave. It must be supposed that she had fallen a victim to her own curiosity. The stiller decided that it was not right to waste valuable mash, so the bloated and perhaps somewhat disintegrated remains were fished out and the distillation proceeded according to plan. Now it will be granted that no hazard to public health was involved. As is well known, alcohol is an almost perfect germicide, and, as an additional safeguard, the heat of distillation would be equivalent to thorough sterilization. It was merely a question of esthetics. However, the grisly secret was not well

The Curious Cat

kept. Somebody babbled, and there ensued a great deal of unfavorable local publicity, though my informant added that there is no record that any of the regular patrons went permanently on the water wagon. Possibly, the incident may have hastened the downfall of this particular enterprise.

In my time, I have talked with ancient tanners, coopers, cobblers, and millers, and I have inquired diligently the details of their vocations, but I am sorry that I never chanced on any one (and there must have been such within my memory) who could have conveyed to me something of the stiller's lore. It may well be that it was regarded as an occult art not to be lightly transmitted to other men. I knew Thomas Wing's son, but I never asked him concerning his father's trade. There must be at least a five-foot shelf of technical volumes dealing with brewing and distilling as carried on today, but the old-time experts left little of tradition and almost nothing of the written word.

I have, however, come on this one bit of writing. Ezra L'Hommedieu was born on Long Island in 1734, and his long life carried him well into the nineteenth century. Along with many political and educational activities, he found time to cultivate a vast enthusiasm for agriculture, and in the primitive journalism of his time he seems to have been one of the very first men to break into print concerning his farm beliefs and practices. In the *Transactions of the Society for the Promotion of Agriculture* (1792–1799) may be found one of his contributions relative to "A Method of Distilling Ardent Spirits from Potatoes." It is a somewhat detailed account of his personal experimentation along this line. He not only reports that his efforts were crowned with success, but he grows positively lyrical concerning the peculiar excellence of his product. He envisions a possible great future for the potato grower as the result of his research. Alcohol can readily be made from almost anything containing starch, and a great variety of materials have been used. However, Ezra to the contrary, in America at least, potatoes have never been deemed a satisfactory source of beverage liquor. In the homespun age, the distiller relied mainly on corn and rye, and less commonly on barley and oats. At an early day (and still) some distilleries, especially in the Hudson Valley, made apple brandy, commonly called apple jack.

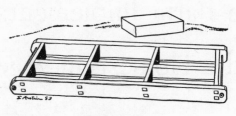

Brickmold

CHAPTER XX

The Farm Implements of the Homespun Age

BY UNIVERSAL consent, the plow is the one implement which above all others typifies the vocation of the husbandman. So the poet Cowley, long ago when heraldry was still an everyday thought in the lives of men, wrote that "a plow on a field arable is the most honorable of ancient arms." The plow is the accepted symbol of agriculture just as the ancient straw beehive is the symbol that stands for thrift and industry.

Our modern field studies of the mechanics of tillage emphasize the fact that for the purpose of soil pulverization the spiral wedge of a properly constructed moldboard is still the peer of all cultural implements, but until about the year 1800 our fields were turned—or more correctly rooted up—by plows which were not greatly in advance of the crooked stick of immemorial centuries.

In 1798 that many-sided genius Thomas Jefferson turned from the cares of state to write a dissertation upon the use of the plow and the proper shape of its moldboard. A year earlier than this, one Thomas Newbold of New Jersey had patented a plow with the moldboard and landside cast in one solid piece. In 1814 and again in 1819, Jethro Wood of Cayuga County, New York, patented the separate cast-iron share. The common walking plow is a wonderful implement. Emphatically it is a tool of most delicate adjustment, and the proper shape of its curves and the relation of the three sides of the triangular wedge involve some considerations which belong to the realm of higher mathematics. Many inventors have worked on these problems and improvements have been rapid. Since 1850, at least, we have had available fairly satisfactory and efficient plows.

I have seen at least two early plows provided with only one handle, a construction that would seem to offer every possible disadvantage. One of these plows was reputedly used in digging the ditch of the first Erie Canal—a work which occupied the years 1817–1825.

The New York State College of Agriculture has a plow the history of which seems to be well authenticated. In the year 1792 the Rappalee family migrated from New Jersey to the then wilderness of Seneca County and in their wagon, together with the other rude belongings of the pioneer, was this plow. It is clumsily but strongly constructed of wood throughout. The handles are made of sticks selected with a natural crook, the moldboard is a massive slab hewn into a rough approximation of the curve of a modern plow, and the different parts are fastened together with wooden pins instead of bolts. A strap of iron protects and strengthens the rear end of the moldboard. Doubtless some bit of iron once shielded the wooden point that acts as a shear, but this has been lost. All in all, it is a pitifully clumsy tool. Our sympathy, and our admiration, goes out to the heroic men who conquered the land with implements such as this. This plow may have had a certain measure of usefulness on the light, stone-free soils of the New Jersey coastal plain, but it seems impossible that it could ever have found much place on the stubborn clay loams of Seneca County.

Probably the first harrows were even more primitive than the plows. Until in very recent years, perhaps even until now, the brush-drag has been found an excellent tool for covering grass seed, and it is a safe guess that a small bushy tree dragged by oxen was the earliest harrow of our pioneers. I do not, however, remember any statement in print that bears this out. On the farm of the writer, a large storage barn was burned about the year 1900 and along with it implements no longer used that went back to the early years of the farm's history. Among them was a drag, shaped like the letter A with straight iron pins for teeth. I dimly remember the drag but very little detail regarding it. My father told me that it had nine teeth and that in his earliest memory, say the 1840's, it was the one and.only harrow on the farm. Many were the long days that the panting oxen dragged it back and forth across the fields.

The gulf between the primitive soil scratcher and a modern tractor-drawn cutaway disk harrow is as wide as between the sickle and the grain binder. Fortunately for the pioneer, though, the problem of soil pulverization was very much less difficult than now. The newly cleared soil was still full of humus and fell apart almost at a touch. A decrease in the amount of decaying vegetable matter has resulted in clods and the need of more efficient cultural implements. Then, too, our pioneer grandsires, whatever their other troubles, were spared a long list of weeds, mostly of European origin, which make life burdensome for us.

The earliest efforts to till the ground between the rows of corn and so control the weeds was by the use of a very small, light, one-horse plow. These were real plows but so small and light that they seemed little more than toys. However such plows were never the right answer to the problem of clean cornfields, and as early as 1840 we had corn cultivators. Various specimens of this class of tools may be seen in the Farmers' Museum at Cooperstown. They are heavy, clumsy devices with wooden beams, but nonetheless, they are unmistakably the progenitors of the one-row, one-horse implements which in the Northeast pretty generally held the field until almost the turn of this century.

The pioneer farmer needed also hand culture tools: hoes, spading forks, potato hooks, and rakes, and these were made by the local smith as he wrought by his charcoal fire Their genuineness is attested by the welds and hammer marks embedded in them. At best they were clumsy, heavy "man killers" which today would cause any self-respecting hired man to throw up his job. It is evident that there was once a large number of these handmade tools upon our farms. That rather amazing repository of early farm Americana, the Farmers' Museum, has many such specimens.

Most of our planting and seeding machinery is relatively modern. The pioneer had so many problems subduing the forests, breaking up the soil, and later harvesting the crop that the mere question of getting seed into the ground appeared a rather minor consideration. As a matter of fact, drills or planters have never represented any such epoch-making advances as are represented by power harvesting and threshing machinery. Before the coming of the silo, corn was seldom handled in large areas in New York State and the "dropping" of corn was regarded as rather specially the chore of small boys and girls and not infrequently of women. I suspect that the youthful laborer, as he carefully counted and dropped the golden nuggets, refreshed his memory and at the same time imbibed sound agricultural practice by repeating the ancient rhyme:

> One for the black bird,
> One for the crow,
> One for the cut worm,
> And three to grow.

Dropping corn thus in the warm May sunshine and covering it with a scrape of the bare foot was in no way a hardship nor was it an unending task.

So, too, the skillful hand-sower could cast seed to cover a

large area in a day, more than could be sown with the usual drill, and some men did this most essential work with remarkable accuracy. Hand-broadcasting of seed grain persisted down to a period when most farm occupations were more or less mechanized. Even after drills had been very generally adopted, it was not unusual to come across elderly farmers who would stoutly insist that the old way was best. The hand-sowing of grass seed in early spring on the winter grains, wheat and rye, was a general practice until quite recent years.

It is not easy to fix a date for the first introduction of seed drills. In the old Albany *Cultivator* of June, 1850, is an advertisement of Sherman, Foster and Co. of Palmyra, Wayne County, calling attention to their new machine for seeding grain and at the same time broadcasting ashes, plaster, or guano, the three standard fertilizers of the time. Moreover, the machine was "warranted" and the price seems very modest, only $65.00. Still this was quite a sum of money and a pretty questionable investment at a time when the world was full of husky farmhands working from sun to sun at a good deal less than a dollar per day.

About the middle of the last century there was a great wave of enthusiasm for the raising of mangel beets as good for cattle and sheep, a practice whose only drawback was the great amount of hand labor required. A drill specially adapted for this work was made and one of these was once used on the farm of the writer. The date of this particular implement can be definitely fixed. In the *Cultivator* for April, 1850, is an advertisement of H. L. Emery, Agricultural Warehouse and Seed Store, 369–371 Broadway, Albany, New York. This firm, by the way, was a regular advertiser in the agricultural papers of the period and seems to have been a pioneer in the distribution of early agricultural implements. The advertisement referred to carries an unmistakable picture. The implement is stated to have been in use for four years, which would make the date of its introduction 1846. There is little doubt that it was one of the earliest attempts to replace hand planting of seed by mechanical devices. The machine was recommended for all small seeds and also for peas, beans, and corn, either in drills or hills. The price was only $14.00, a surprisingly small amount.

It is often difficult to assign anything more than an indefinite date for the introduction of an implement. In the Farmers' Museum there are a number of drills such as might be used for sowing garden seeds and also corn planters of the type still in use within the memory of elderly folk. Their construction, largely of wood, with iron used only where absolutely essential, indicates that they belong to a relatively early period. There is nothing, however, that will enable one to say precisely in such and such a year, "This tool first came into use." The most hopeful method of determining the answer to questions of this sort is by a systematic reading of the advertising pages of the few agricultural journals of those years along with a diligent perusal of the *Transactions* of the New York State Agricultural Society, these *Transactions* having been over a period of many years the favorite medium through which the experience and practices of the best farmers of that day were given to the world.

It is not easy to determine just what implement can be regarded as the grandfather of the cradle. I am pretty sure that it is not the sickle of the familiar type we see as part of the paraphernalia of Grange halls or in illustrations symbolically crossed with a sheaf of grain. I do not know of any evidence that this sickle was commonly used as a harvesting tool in America, though it is certain that there were implements which antedated the cradle. My father sometimes spoke of the "sith" once used on this farm, but I cannot be sure if he spoke from actual memory or from oral tradition handed down to him. As I remember it, the use of this tool was linked with a certain farmhand, the redoubtable John Brown, who had other methods that were different from those of most people. In any event, this bygone and almost forgotten tool was neither a sickle nor a cradle nor yet a scythe, but an awkward, angular contrivance very different from any of the foregoing. It was wielded with the right hand while the "mathook" in the left hand caught and drew the standing grain within reach of the blade. As the grain was cut off, a deft sweep and toss of the hook threw it into a swath, or windrow, on the left of the operator. Such at least is the tradition regarding the purpose of the sith, although I doubt if there is today any man living who could give authentic demonstration of its use.

The spelling of the word "sith" is strictly phonetic and my own. In the *Century Dictionary*, "sithe" is given as "obsolete for scythe," but there is no indication that a similar word was used for what was definitely a different tool, which might imply that there are bits of lore which even the big dictionaries miss. In any case the sith was a very early tool and yet it must once have been very common. There are, even now, a considerable number of survivals. In Schoharie County, I have seen the implement among the relics of the Yankee as well as the Dutch and German families. At Cooperstown are several specimens, and the handles worn and polished by the grip of faithful men, now long dead, attest that they have seen years of service. The sith seems to me an honorable ancestor of the cradle, although I do not propose to be entangled in any controversy as to the precise number of generations removed.

I do not see that it is possible to attempt any historical sketch of the evolution of the cradle. Professor William Henry Brewer (1828–1910), Yale Sheffield Scientific School, states very positively that it was a New England invention and even fixes a definite date, 1776, for its appearance, but he does not give the name of the inventor. That great teacher of agriculture at Cornell, Isaac Phelps Roberts, in lecturing to his classes, was fond of emphasizing the fact that the cradle constituted an epoch-making advance over all former harvesting implements and declared it to be a greater stroke of absolute genius than the self-rake reaper or the binder. In any case the tool seems to have won speedy and almost universal adoption wherever grain was grown in America, and it came to be widely used in Europe as well.

It is certain that at least a century ago, the cradle was the almost universal implement of harvest, nor is there any suggestion that it was then a new invention. Aside from its scythe, a piece of forging not beyond the skill of the old-time smith, the material for its construction was abundant in every community. True, to produce a cradle called for skilled craftsmanship in wood, but without doubt skilled men were once more common than in this age of automatic machinery. The all-day use of the cradle called for mighty biceps and an unbreakable back, but these were qualities not lacking in our pioneer ancestors. There were men who prided themselves on their ability to cradle all the long day with the tireless swing of a steel mechanism. These mighty old-time champions (as well they might be called) were very fastidious as to their favorite tool and so we find the implement constructed in various models. Doubtless, there were many of these befitting the whim of the maker or owner, for standardization is a product of machine method. Old men speak of the "straight," the "grapevine," the "half grapevine," and the "mooley."

Only slowly did the cradle pass from our farms. In 1832 Obed Hussey, of Baltimore, was granted a patent for a reaping machine and a year later Cyrus McCormick of Virginia took out a patent for a similar invention; but in New York, at least, the cradle substantially held the field until after the Civil War. Indeed, I do not feel that the reaper was really common in our state until as late as 1875, but this is a rather arbitrary date to fix the coming of a very gradual evolution. As a matter of fact, the cutting of grain with a cradle was by no means as slow a process as may today be imagined. When grain stood up well and was not lodged or tangled, a good cradler was able to lay a surprising acreage in a day.

It is interesting and historically important to determine just what constituted a good day's work in cradling grain. Through the courtesy of the columns of the *American Agriculturist* I broadcast this inquiry, addressing it primarily to aged farmers who might have this knowledge either from direct experience or from reliable tradition. My inquiry brought me a goodly number of replies, written, in many cases, with most painstaking care and certainly representing the usual estimate of the ability of an old-time cradler. On the whole there is a surprising agreement of opinion. It seems to be generally accepted that under favorable conditions a good man ought to cut four acres of grain in a day, and more of oats and buckwheat than of wheat or rye. There were a few frankly big stories, running up to seven acres or more. Of course, men always remember and recount the big day's work and not the unsuccessful one. Nevertheless, I believe that four acres of grain between sun and sun was by no means an uncommon feat for one of these bygone athletes. Perhaps it would be safer to discount these estimates by one-third to one-half for the usual everyday working basis.

Mowing grass with a scythe was a much slower job. Perhaps, on the average, hardly more than one-half the acreage was covered than was possible when cradling. The man with the scythe must take a narrower swath, and he could not reach so far forward at each clip. The cradle was a heavier tool to swing, but the grain being cut with a long stubble required very much less power. Many men declared that they enjoyed cradling, but mowing was by common consent grueling. Cradling performed by a good man was a beautiful exhibition. A band of well-matched cradlers going down a field of golden grain had all the rhythmic, measured swing of a college eight-oared crew. Except in memory, we can never see that spectacle again.

Up until within the memory of many living men, our grain was not only cut with cradles but bound by hand as well. Indeed the invasion of the Canada thistle, together with the growing difficulty of securing labor, was the cause that led to the general discontinuance of the practice. To rake a sheaf of grain from the cradle was an awkward operation, but it was one where long practice gave a strange and almost automatic dexterity. An old-time familiar and oft-repeated feat was to bind one sheaf, toss it high in the air, and then bind another before the first touched the ground. In the early days this binding was frequently the work of women. In my very faintest boyhood recollections, on a neighboring farm lived one Austin Stahl, a farmer of a pioneer line. According to neighborhood tradition his mother was the acknowledged champion among the women binders of her day. My father repeated to me the legends of the exploits of Mrs. Stahl—how in the strength and suppleness of her young wifehood there were very few men who could equal her in binding grain; and, more specifically, how she would bind behind a cradler all day long, pressing close at his heels, and at the end of the day in a spirit of mockery would catch in her waiting arms the last clip off the cradle and bind it without allowing it to touch the ground.

There is every reason to suppose that the practice of binding grain goes back to the very beginnings of husbandry and the childhood of the race. It is written in books how the old Greeks embossed their battleshields with many scenes, of battle, of love, and of dancing; of cattle under great trees, of men plowing with oxen, and of reapers binding sheaves with the master standing in their midst watching their labors. So, too, I remember the story of the ruddy lad Joseph and his dream of how he and all his turbulent brothers were binding sheaves upon the Judean hills, and then his foolish tale of how his sheaf arose and stood upright and the sheaves of all his brethren stood round about and did obeisance to it.

Perhaps in all the world there is no more ancient discovery than the binder's knot, that deft twist and tuck which the expert accomplishes so rapidly that the eye can scarcely follow the hand. I have read how the Swiss lake dwellers lived so long ago that no one ventures to establish a date for their history save that it was before the written record or even oral tradition. Yet dredging in these lakes has brought to light fishing nets tied with the identical knot employed today by deep-sea fishermen throughout Europe, and doubtless the same that was used by those two never-to-be-forgotten fishermen who one far-off day very suddenly "forsook their nets and followed Him." So I love to please my fancy by thinking that the sheaves of wheat that my great-grandfather bound on Hillside Farm were in no way different from those made by the reapers behind whom Ruth gleaned in the field of Boaz.

In our early New York state agriculture, wheat and rye, at least, were almost universally bound. Frequently, perhaps generally, this was true of oats as well. I can rather dimly remember how in my earliest boyhood—it must have been almost seventy-years ago—our neighbor, Asa Abbott, bound by hand and beautifully shocked up, every shock protected by a cap-sheaf, several acres of oats. I think this must have been the last of this sort of work in our community.

To bind grain was ordinarily a bigger job than to cradle it; so it is not surprising to find early attempts to reduce this labor. There is an interesting woodcut of a "grain binding wheel-rake" in the *Cultivator* for August, 1850. The way in which this tool was to be used is self-evident. The advertisement states: "It is wide enough to handle the longest straw and is light, weighting only about 15 pounds." It was to be pushed along the cradle swath ahead of the operator until it had gathered enough for a proper sized sheaf; then the pressure of the foot on a pedal would tilt the bundle up into position so that it could be bound without bending the back. Like everything else of that period the price was low, from three to four dollars. It was surely an interesting and ingenious device and I have no doubt that it could be worked, but there is no evidence that it ever attained any general use. Even so, it is not the only invention that never got beyond the working model stage. As a matter of fact, the old-time binder was so dexterous and so nearly automatic in his action that any contrivance of this sort would be only a hindrance so far as speed was concerned.

That same year (1850) the *Cultivator* published a sort of pictorial supplement which was, without doubt, the most ambitious attempt at illustrations to be found in any agricultural publication up to that time. There are twenty-eight pages closely filled with quaint woodcuts of the animals, implements, and devices of that day. Unfortunately, there is hardly a word of explanation regarding these pictures. One of the most interesting is Easterly's harvesting machine. I very much question if it ever got beyond the patent model stage, though it is plainly the ancestor of the big combined header and thresher used in certain wheat-growing districts of the world.

McCormick was granted a patent for a reaping machine in 1833 and again in 1845 for a later form of the same machine. This particular type of reaper came to be rather widely used. It required a second man who rode on the machine and with his rake swept a gavel from the table whenever enough grain had accumulated to make a proper sized sheaf. Of course the next, and very obvious step, was to replace this human raker with an automatic device as part of the machine—hence the term "self-rake" reaper. Still, a perusal of the agricultural journals of the period would not indicate that the reaper found any extensive use in our state previous to the Civil War. After the introduction of the reaper came that wonderful combination of reels, elevators, and steel fingers, the binder, using at first wire but a little later twine. On our New York State farms this modern binder with bundle carrier (sometimes tractor drawn) remained the last word in harvesting machinery until crowded out by the combine.

New York has had a number of different crops that were once important but now are almost forgotten. One of them was the Canada field pea. In 1844 the state grew more than 117,000 acres of this crop, all for grain, because the canning factory crop had not yet been dreamed of. All of this big acreage was cut, not with

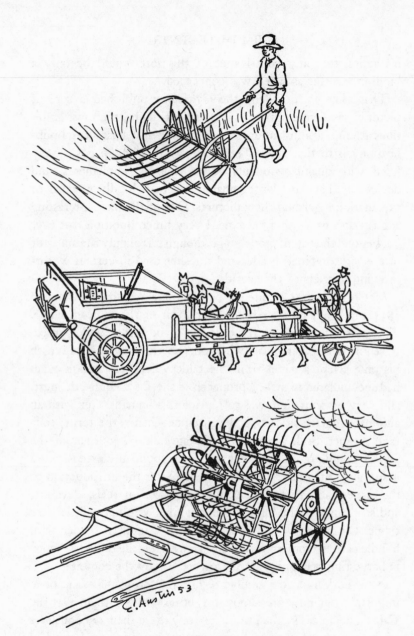

Reading from top to bottom:
Wheel-Rake, Easterley's Harvester, Hay Spreader

cradle or reaper, but with the scythe. The mower did not take a regular swath, but worked into the field from the edge and "rolled out" the pea straw in bundles about as big as would cure nicely and make a handy forkful. My father often assured me that there was nothing in the harvest line half as much fun as pitching these neat forkfuls of peas when dry and ready for the barn. On one occasion John Brown pitched nineteen big loads of peas after dinner, and this feat, by much retelling, gradually became a legend. I shall rehearse it to my grandchildren as another incident of the farm Odyssey.

The twine binder marked an epoch in agricultural economics. The era of disastrously low prices which overtook agricultural products from about 1875 until the end of the century had several different causes. Most of all, perhaps, it was due to the opening of the vast virgin empire of the trans-Mississippi states, an addition to the world's agricultural resources so stupendous that there was no possibility of the rapid absorption of its products at any remunerative price. Other causes of the price decline were improved methods of steel manufacture and cheap rails, bringing an era of railroad expansion such as had never been known before and which can never be repeated. But, along with these causes, due weight must be given to the invention of an automatic machine that could cut and gather a sheaf of grain, tie it with a bit of string, and then toss it aside. This machine made possible an expansion of cereal production such as could not otherwise have taken place, and it is not too much to say that if this invention had never come to pass, the eastern farmer might have been spared many of the hardships of "the great agricultural depression."

It will be seen that our grain harvesting machinery is the product of a long evolution, every step of which seemed an almost infinite advance over the previous stage. Various modifications of the immemorial reaping hook provided the implement with which were harvested the first narrow fields of the American pioneer. Later some forgotten genius took his scythe and added to it a light elastic wooden framework with fingers parallel to the blade, and this device enabled him to cut his grain and leave it lying smooth and unbroken and ready for the binder. Hussey and McCormick were only the best remembered of many inventors who labored and dreamed through many years to make horsepower take the place of human muscle.

The cutting of grass is not so long a story. Just as far back as the antiquarian can go the husbandman had forged a curved steel blade, mounted it upon a crooked stick (the snath), and with it swept his meadow lands. The earliest pictures of English agriculture show scythes in no essential respect different from our own. Indeed, one of the encyclopedias hazards the statement that the scythe remained practically unchanged for at least ten centuries. I have also found the statement that not until the seventeenth century was the scythe blade stiffened and improved by the addition of a rib along the back. It was said, too, that the eariest snaths had only one nib, or short holding pin, instead of two as in our time. It will be noted that most cradle snaths follow the earlier fashion. This indispensable tool was among the very earliest objects of manufacture in America. In 1646, only twenty-six years after the Pilgrims first set foot on Plymouth Rock, the General Assembly of the Colony of Massachusetts Bay granted to Joseph Jenckes of Lynn, for a period of fourteen years, the exclusive right to make scythes in the colony. His work and conduct apparently proved satisfactory because the privilege was later extended for an additional seven years. New England seems to have long remained the center of the industry, scythes being forged in literally scores of little shops. Yet as early as 1812 the firm of S. & A. Walters of Amsterdam, Montgomery County, New York, produced 6,000 scythes annually.

The cradle may be about to pass but the scythe will always have a place on the farms of men who believe in clean fence rows and in mowing out the corners of the meadows. My father lived long enough ago to remember when each year at least one hundred tons of hay were cut on this farm with scythes. I wish that he might tell me once more how he had seen as many as seven scythemen going down the field in close array with Willis Goodyear leading the van.

These bygone men contrived to make of the heaviest and most exhausting labor an athletic contest. At the end of the field they paused to breathe and wipe the sweat and take a great pull at the water jug. Then they proceeded to put a razor edge upon their blades, and in all rural sounds there is none sweeter and more musical than the rapid swish of the whetstone on the ringing steel. Always they laughed and gossiped and chaffed a little. Then the man whose turn it was to lead struck three smart taps with his stone upon his scythe, a sound that was both a signal and a challenge, and they were off. If someone lagged in his stroke the fellow mowing literally at his heels cried out the jocular warning: "Get out of my way or I'll cut your legs off." In such fashion our New York State meadows were mowed until a period within the memory of a few of our oldest living men.

Close mowing demanded a stone-free meadow. A single sweep across a stone was a serious disaster to the keen blade. The stones were picked with a scrupulous care that is now absolutely forgotten. This tradition lingered for some time after the necessity of doing so had passed. I can remember the time when we picked the stones from the meadows, each man carrying a battered pail and an old-fashioned gutterhook with which a small half-buried stone could be loosened and lifted from its bed. Stones were gathered literally down to the size of hens' eggs. It was a type of work possible only under the old-time unlimited abundance and cheapness of farm labor. I set it down ungrudgingly that my forefathers were far more exacting and painstaking farmers than I shall ever find it possible to be.

The germinal idea of a horse-drawn mowing machine antedated its successful development by a good many years. As early as

1803 a patent for such a device was granted to two men in New Jersey, and during the next thirty years several other patents to different individuals, including one to Jere (probably Jeremiah) Bailey of Pennsylvania for a machine whose essential mechanism was a vertical shaft revolving and carrying six scythe-like blades. Various other similar devices show that it was very hard for these early inventors to get away from the idea of a sweeping scythe. All of these various inventions died stillborn and one William Manny, of New Jersey, is generally credited with having built the first mowing machine that would really mow. This was about 1831 and his machine is said to have embraced the essential features of the mower of today. Many inventors tried their hands at the problem and small establishments sprang up in many parts of the country. In July of 1857 a very noteworthy trial was held in Syracuse under the auspices of the New York State Agricultural Society. On this occasion some fifteen machines competed and a number of medals and diplomas were awarded. With the sole exception of McCormick the very names of these manufacturers are all unknown to this generation.

Many years ago, before an audience made up of the students of our State College of Agriculture I heard that wise Master-Farmer, the late J. S. Woodward of Lockport, declare the greatest boon that invention ever brought to the farmer in his time was the wooden "flop-over" horse rake. This honorable implement is said to have been invented by Moses Pennock of Pennsylvania in 1824. It attained a very wide, it might be said almost universal, use, and lingered long. I cannot find any hint of just when the modern horse rake was introduced. A farmer writing in 1865 says: "The fine-toothed sulky rake is superseding the wooden rake on the largest farms," but, nevertheless, he seemed to question their advantage. In any case the "flop-over" was an almost immeasurable advance over the wooden hand rake heretofore used.

We usually think of the hay tedder as a machine of rather recent introduction, but a hay spreader (not tedder) is figured in the *Cultivator* of 1850. I can find no advertisement of its sale, however, nor any reference to its use.

The pitchfork remains one of the hand tools that no inven-

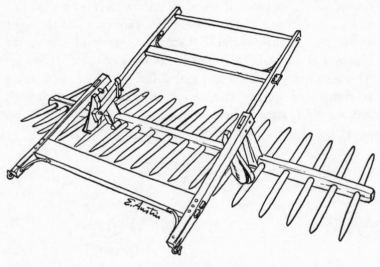

Floprake

tion can ever wholly replace, and I suppose that unborn generations of men will grow hornyhanded and weary in its use. Perhaps there has been no other one article that so well illustrated the absolutely primitive art of the pioneer as did a wooden pitchfork once owned by the Frederick family of Schoharie County. This was nothing more or less than a sapling with three forking branches, fortunately so placed as to form the tines. There was in its make-up no scrap of iron nor the mark of any tool other than the ax. I suppose it was with forks like this, neither better nor worse, that men have handled the harvest for a thousand years. The worn and highly polished handle gave incontestable evidence of much service through long years. The owners stated that it was made (perhaps it would be more accurate to say "cut down" or "selected") in 1802, although I confess I do not know how they arrived at this precise date.

A somewhat later fork is a sapling having the two outer tines formed by the branches, but the maker had inserted a center tine of a separate piece of wood, requiring therefor a small bolt and a screw, thus adding the art of the carpenter and the smith to the craft of the woodsman. Later forks were all of iron, but clumsy and heavy, being evidently the work of some blacksmith. The beautifully polished and gilded fork of today, with every refinement of weight, curve of handle, and tines, apparently leaves nothing more to be desired.

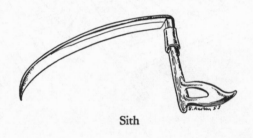

Sith

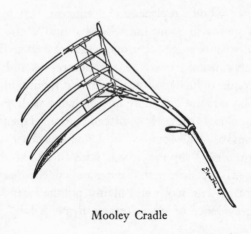

Mooley Cradle

CHAPTER XXI

The Ways by Which Our Fathers Threshed

WITHOUT a doubt the earliest effort to lighten the drudgery of the human arm in threshing was by driving draft animals over the sheaves, which had been unbound and spread in a layer on the floor. This was the universal method which held through centuries and which persisted in our midst until within a generation. The prohibition of the ancient Mosaic code, "Thou shalt not muzzle the ox when he treadeth out the corn" (Deuteronomy 25:4), is at once the world's oldest bit of humane legislation and a testimonial to the exceeding antiquity of the practice. In lands blessed with a rainless harvest climate, the threshing floors were made of smooth hard-trodden clay, uncovered and were situated on hilltops to secure favoring breezes for winnowing. In New York State the necessity for a proper threshing floor was a determining factor in our barn architecture. Very many of our old-time barns were framed with one or two giant "swing beams"—timbers of unbelievable size, sometimes two feet or more square. These were designed to carry the

239

big overhead mows without any center support, thus leaving a clear, unobstructed threshing floor. In the middle of this a small post was set up and a light sweep attached to serve as a guide for the horses that were driven in a circle around it—often with merriment and the crack of whip. Meanwhile, men kept turning over the straw and shaking out the grain and chaff. It was planned to do this work when the horses were without shoes and it was deemed an especially good opportunity to accustom colts to handling. Oxen were also sometimes used in the same way.

Later someone devised a "groundhog thresher" (one of its several names), which was a log about ten feet long into which were driven many stout wooden pegs. One end was pivoted to a post, while a team tied to the free end drew it around and around. As the team made their rounds, the log bumped and rolled behind them, and doubtless it greatly increased their efficiency. I think I saw the thresher in operation once, just once, in my life, but it was so long ago and the recollection is so shadowy that it seems only a hereditary memory, if such a thing is possible.

The flail lives vividly in pastoral literature, but scarcely at all in agricultural writing. Neither in our old agricultural books nor in our early agricultural journals can I find anything beyond the most passing references to this ancient and well-nigh universal implement. Probably the reason lies in the fact that the flail is so exceedingly simple in construction that it hardly seemed worthy of discussion. Essentially, it consisted of a light, elastic handle to which a somewhat heavier and shorter stick (the swingle) was tied by a tough thong. This thong wore out rapidly and was the one weak point in the contrivance. The most indestructible material ever found for this purpose was eel skin and I can remember when two or three eel skins were nailed up in the barn ready for use. The skins remained for years after the flails had been laid away forever. Some carefully made flails

Groundhog Thresher

had a wooden swivel attached to the handle in a way to prevent twisting or kinking of the thong.

So far as I can learn, oats, barley, and buckwheat were generally threshed under the feet of horses; but wheat, rye, and peas were flailed. During the first half of the last century, a considerable acreage of wheat was grown on my family's farm each year, and until about 1850 it was beaten out with flails. Wheat is hard to thresh and needs pounding to clean the grain from the straw. The wheat must be fully ripened, and sometimes it was exposed to rain before drawing in, though I never heard this particular bit of farm practice from any one except my father.

The use of the flail for threshing rye was kept up long after it had been discontinued for other purposes. For one thing, rye threshes more easily than any other grain. More important was the fact that rye straw had a special sales value for making old-fashioned brown paper and for other uses, but this value was conditional upon the straw being straight and unbroken and bound in bundles. The sheaves of rye were unbound and laid on the floor in two rows, the heads overlapping and the butts outward. Then two flailsmen worked systematically over them, striking alternately in the same spot and beating a rhythmic tattoo. The straw was then turned over so as to present a fresh surface and the process repeated until the grain was beaten out. When finished, the chaff and grain were carefully shaken out and the straw, still straight and unbroken, was rebound by hand, a process that to us seems almost incredibly laborious.

I can dimly remember when this was done on our farm. I have before referred to John Brown, that fine-hearted Irishman who for sixty years spent most of his time with us, and who was finally gathered to his fathers at the age of ninety "at a good old age like a sheaf of corn ripe for the harvest." He and another worthy, James Barker, used to work together and flail out rye "for the tenth bushel." I am told that ten bushels per day per man was a busy day's work, and rye was commonly worth about seventy-five cents per bushel; but this amount was counted a satisfactory wage for a winter's day a century ago. Grain flailed easiest on snapping cold winter days, but it was work at which a man had no trouble to keep warm and it was regarded as pleasant employment for rough weather.

Later the "rye-beater" came in, a special type of threshing machine with a very long cylinder through which the straw could be passed lying lengthwise, and thus threshed without tangling or breaking. A man, sometimes two men, stood behind the beater and rebound the straw as it was delivered on a table in front of them—a job specifically recognized as calling for quick and skillful workmanship. Later still, some man made the obvious improvement of attaching a twine binder to the rear of the beater, and with this disappeared the last stand of the hand binder and the harvester's knot. And we shall never again on winter days hear the measured, muffled beat of flail on threshing floor.

But whether the threshing of the pioneer was accomplished by trampling with animals or by flailing, getting the grain out of the straw was only half his task. There remained the winnowing. It seems to be generally agreed that among all peoples and through all centuries this was accomplished by the aid of favoring breezes. Thus, Daniel the prophet, when he wished to convey the idea of the utter destruction and dispersal of the image with its head of gold and feet of iron and clay writes that the pieces "became like the chaff of the summer threshingfloors; and the wind carried them away" (Daniel 2:35). There is no question that along with the rest of the world, the American pioneer cleaned his grain by pouring it or tossing it in a favorable current of wind. He made himself a special implement for the purpose: a very large, flat, shallow basket which could be filled with chaff and grain, then lifted up, and the mixture poured in a thin stream over the edge. I am sure that this basket must have had a special name, but I cannot learn just what that name was. There seems to be a considerable number of these baskets still in existence, and I assume that they were once to be found on all good farms, but they passed from use so long ago that I find it impossible to assign any date. My friend and neighbor John Shafer assures me that he remembers his grandmother and grandfather standing on the barn floor with the big doors open on each side so that wind might have a clean sweep. They would fill their basket with chaff and grain and then, one at each end, would lift it high in the air and shake the contents gradually over the edge. The wind did the rest. He adds that the process was surprisingly rapid—a statement that is easy to believe for conditions of steady and favorable wind.

Nevertheless, men sometimes tired of waiting for a fair wind. Long ago some primitive mechanical help was brought into use.

Thus John the Baptist, preaching in the desert beyond the Jordan, picks up a familiar rural illustration and cries: "Whose fan is in his hand, and he will thoroughly purge his floor, and gather his wheat into the garner" (Matthew 3:12). We have had fanning mills for a great many years and the fanning mill of 1840 looks very much like the mill of today. There have always been well-meaning souls who for one reason or another have arrayed themselves against all innovations, and so in the early days of the last century we find a hard-boiled Scotsman protesting in the *Cultivator* against the use of the fanning mill, basing his reasons on theological grounds, thus: "Your Ladyship and the Stewart have been pleased to propose that my son Cuddie should work in the barn wi' a new fangled machine for dighting the corn from the chaff, thus impiously thwarting the will of Divine Providence by raising the wind for your Ladyship's own particular use by human art."

The desirability of some sort of mechanical device for threshing became evident a great while ago, and as early as 1732 there is reference to such a machine in Scotland, "where bye one man may do as much work as six men heretofore," and from time to time during that century other men tried their hands at the problem. In any invention the real stroke of genius is the grand original idea. After that, all else is easy. Just as for many years the inventors who worked at the mowing machine sought to imitate the action of shears or the sweeping motion of the swinging scythe, so the first threshing machines were built on the principle of rods which rose and fell and beat out the grain after the plan of the familiar flail. The real invention of the modern thresher took place on the day that a man bethought him of a swiftly revolving drum armed with metal teeth. This epoch-making idea is said to be a strictly American conception. About the year 1822 a New Hampshire Yankee, whose name I cannot discover, migrated to Saratoga County, New York, and brought with him in his wagon, or perhaps in his head, a threshing machine

Reading from top to bottom: Primitive Thresher, Flailers,
Threshing Machine, Wooden Fork, Winnowing Basket

embodying the great idea. A little later the plan was taken up and further developed by T. D. Burrall, a farmer-mechanic of Geneva, N.Y., who about 1830 began to manufacture and market this machine. Thereafter the evolution of the thresher appears to have been surprisingly rapid.

These first machines were nothing more than a cylinder and concave, without any provision for separating the straw from the chaff and grain; but compared with former methods, they were a revolutionary advance. The first source of power was the sweep-power with from two to six horses, but about 1840 the tread horsepower came in. I cannot fix a precise date for this but I can find no reference to their use before 1845. These early machines simply discharged a mixture of straw, chaff, and grain into one common heap, and it was no small job to shake out the chaff from the grain. A machine of this very primitive type was used as an illustration in a periodical of 1844. One like it was used on our farm within my father's memory. A few years later a monumental advance was made when some genius added a "shaker," which allowed the grain and chaff to be shaken out of the straw and to fall beneath, while the straw was carried off and could be handled separately. This simple type of machine, with only a cylinder and shaker attachment, lingered on many eastern New York farms for two generations. As a boy I had experience (altogether too much experience it seemed to me) with just such a rig. Indeed, I know a farm where a machine like this was still in active use in the winter of 1926. Because of their extreme simplicity and the absence of all extra fanning or elevating devices to consume power, two good horses would thresh a surprising amount in a day.

Improvements were introduced very rapidly. In 1833 it was reported that a machine was used in South Carolina that would both thresh and clean rice in one operation, and as early as 1844 we had a machine that had all the essential features of a modern giant thresher, and yet it was only twenty-two years from the time the forgotten Yankee set up the first thresher in Saratoga County.

In the early numbers of the *Cultivator* and through several years appear occasional contributions from one whose sense of modesty, or of mystery, impelled him to conceal his identity under the rather classical *nom de plume* of "Agricola." His residence was Seneca County and he was unquestionably a real farmer. Moreover, his writings gave indisputable evidence that he was a scholar and a gentleman in the finest sense of the phrase. In 1849 he wrote at length concerning the advantages of the then new machine which both threshed and cleaned the wheat. I quote verbatim his cost account for the threshing, as an illuminating comment on a period when there was surely no farm labor shortage.

One man to feed the machine at		39	cents per day.
do	to supply the feeder	38	cents per day.
do	to pitch from the mow	34½	cents per day.
do	to deliver the straw	32	cents per day.
do	to attend fanning mill generally done by self	50	cents per day.
	Per day	$1.93½	for labor.
	Four horses and driver	2.50	per day.
		$4.43½	per day.

Wheat is threshed at an easy pace, delivering 200 bushels (often more) per day which at the above rates makes the cost 2.0021 cents per bushel. It should be remarked that the rate of wages above named are the actual rates paid, the work being done by yearly hands whose wages amount per day to the sum stated.

I expect that our farm management experts would insist upon criticizing and dissecting these figures, but I do not propose to play the ungracious carping questioner.

"Agricola," although evidently something of the county squire and a man of mature and confident culture, did not scorn to work with his men at harvest and to vote himself the inflated wage of fifty cents per day. It is especially interesting to note that this same month "Genesee wheat" in New York, shipped down, of course, over the old Erie Canal, was quoted at $1.23 to $1.25 per bushel. Ohio wheat, the only other grade quoted, sold for $1.03. I am not sure but that this period is in some ways the Golden Age of the agricultural history of New York.

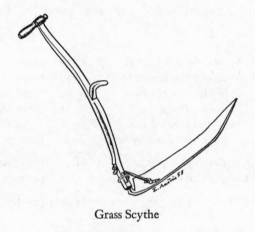

Grass Scythe

CHAPTER XXII

The Household Handicrafts

And all the women that were wise hearted did spin with their hands, and brought that which they had spun, both of blue, and of purple, and of scarlet, and of fine linen. (Exodus 35:25)

I SUPPOSE that the most important of the household handicrafts, was the textile art as expressed in wool and linen. There was a day when a plot of flax was found on every well-ordered farm, just as there was a wheel for flax and another for wool and a big "barn-frame" loom in every kitchen. Flax was grown in the colony of New Netherland as early as 1626 and always thereafter had a most important place in the farm economy of the pioneer.

Well have the botanists chosen for flax the scientific name *Linum usitatissimum*, which means, most useful. As a source and aid to human progress it must be enumerated among the half-dozen most important plants in all the world. Fortunately for mankind it seems to thrive under very wide variations of climate and is successfully grown from the tropics to well toward the northern limits of agriculture from Scandinavia to Manitoba. It is in the cooler climates, however, that it attains the highest quality of fiber. Its use is one of the old, old discoveries of the race. Dredging on the site of the homes of the Swiss lake dwellers brings up fragments of their fishing nets, and skilled microscopists establish the fact that these were made of the indispensable flax. The Mediterranean basin was the cradle of civilization and there the flax plant grew abundantly. It furnished the winding sheet for the kings of the men who piled up the pyramids. The Assyrian and the Greek went clad in the same universal fiber that made the shirt and the kirtle of the American pioneer.

Flax is the premier textile plant of the world. Other cheaper and far inferior vegetable fibers have very largely displaced it for the commonplace uses of life; but when men want strength and durability, when they want beauty, as in table linen and in altar cloths, then they turn back again to this imperial plant. Flax fiber when first prepared, spun, and woven may be of varying colors, but always some soft and lovely shade of gray. Our grandmothers bleached it in the sun until it became almost white, practically without loss of strength. Modern bleaching with boiling and chemicals attain gleaming whiteness but at the cost of durability. Compared with flax, cotton is at best weak and shortlived. Cotton fabrics grow rotten and weak with the passing years but linen literally survives the centuries. For sheer strength a slender thread of twisted flax fibers is incomparable. Woven alone and made into shirt or dress or sheet it made a fabric literally an heirloom. Used as warp to protect and hold the honest woolen woof, it made clothing that kept out the cold or bedding beneath which the pioneer in his rude cabin home might lie warm. In 1845 we had more than 46,000 acres of flax, mostly in small areas.

It is doubtful if the busiest farmstead would require for its own use more than an acre, but if there was a surplus, dressed flax was one of the few commodities that enjoyed a dependable cash market. Ten years later this acreage had fallen to less than 13,000, testimony of how rapidly household spinning and weaving were going out. Under the stress of the Civil War, with famine prices for cotton and probably the patriotic appeal for national self-sufficiency, the dying art flared up again, so that in 1864 we grew more than 18,000 acres, but it was the last stand of the flax wheel. Never again did the census enumerator find any considerable area of flax, and for fifty years the crop has been practically extinct in our state. Indeed, I think it very probable that flax production reached its high-water mark even before the first crop census. It is interesting to record that one or two of our state hospitals for the insane still grow a little plot of flax because its preparation and spinning afford interesting occupation for their patients.

I have sought from many correspondents firsthand information concerning this old-time crop and details as to its culture, preparation, and use. Many old men will remember the crop as a part of their youth, but there seems to be a lack of definite clean-cut memories. All agree to one statement: the crop disappeared soon after the close of the Civil War. There were, however, several articles published in the *Cultivator* and the *American Agriculturist* during the decade 1840–1850 relative to the raising of flax, and these were by men who had had experience with the matter on which they wrote. In the *Cultivator* for 1846 is a communication from one who hides his identity under the simple initial "B." He writes from Leyden, N.Y., and I assume that his experience represents a successful crop grown according to the accepted methods of that time. He sowed one acre of ground which had been stubble the previous year, using three pecks of seed. His harvest consisted of 15 bushels of seed, which he estimates as being worth one dollar per bushel, and 250 pounds of dressed flax fiber, worth $18.75. He places the cost of dressing the fiber at $6.25 and other labor, $2.00, and so estimates for himself the satisfactory gross profit of $25.50. It should be noted, however, that the best quality of fiber was not secured when the crop was allowed to ripen its seed.

Flax was sown as early in the spring as the ground could be well prepared and it was especially desirable that the land be free from weeds, not only because they interfered with growth of the flax but even more because they were a great bother in subsequent handling. For this reason, recently cleared and newly burned-over land was sought, and in the days when flax was an important crop, such land was usually available. The plant grew from two to three feet tall and was ready to harvest in late July or early August. It was always "pulled," not cut, probably because this method permitted it to be put into the best condition for handling and also because longer fiber could be saved. Any breaking or tangling of the straw increased the labor of preparation and resulted in a less desirable product. The finest and silkiest fiber was secured by thick seeding and harvesting while still immature. Flax is a hollow-stemmed plant with long and wonderfully strong fibers, or filaments, which run the entire length of the stalk. The flax fiber, however, forms only a small part of the total weight. The part of the plant that is laid in and around the fibers has no textile value and must be disposed of before the flax can be prepared. This was done by retting, that is, by exposing it to water in running streams, or far more commonly, by laying it out on the grass in a thin layer until partially rotted. The two methods were called, respectively, water retting and dew retting. Water retting was much quicker, being completed in from eight to twenty days, depending upon the maturity of the plant and the temperature of the water. It also gave better control of conditions, and this is the method used now in countries where flax is grown commercially. Probably few farmers were fortunate enough to have a proper stream or water supply available.

In dew retting, the length of exposure varied with the weather and the maturity of the plant but usually occupied several weeks.

As one correspondent puts it, "It laid out until it looked as if it was thoroughly spoiled." This was for the purpose of letting the rotting go far enough so that the interfibrous portion of the plant partially decayed while the true fiber remained uninjured. There were certain rule-of-thumb tests by which to determine when the process was completed. When the right stage was reached the retted flax was gathered up and thoroughly dried. On certain of the best-equipped farms there was a fire-heated loft or kiln for this special purpose.

The flax was then, a single handful at a time, broken on that primitive and universal implement, the flax brake or "crackle." This was essentially a wooden beam four or five feet long, supported, sawhorse fashion, at a height convenient for the operator. On top of this was a second beam, hinged to the first at one end, and so arranged that the other end could be lifted and dropped by handpower. The handful of flax straw was vigorously pounded between the two beams until the boon, or the non-fibrous portion of the plant, was thoroughly crushed and loosened from the true flax fiber. Then it was swingled ("switcheled") by beating it with a great wooden knife along the edge of a plank, thus removing most of the broken waste; and finally it was hatcheled (always a handful at a time) by drawing it over and through the many-toothed hatchel, thus cleaning and combing it out into a beautiful smooth strand of soft gray fiber, the dressed flax. Hatchels of different sizes and fineness were used, the small close-toothed ones producing a finer and more beautiful product. Judged by our standards of industry, the process must have been almost infinitely laborious. The short and broken fiber that was not so well cleaned was called tow, and was very much less valuable than the long clean fiber of the dressed flax. However, it was roughly spun and found a place as filling or woof in the poorer fabrics, the warp being supplied by the long flax. A chance sentence from an early writer in the *Cultivator* indicates that, when handled according to the usual methods of those times, the yield of tow was commonly about one-half the weight of the dressed flax.

The per acre yield of the fiber varied as widely as any other crop. The figures of the 1845 census for the total product of flax fiber, divided by the total acreage, indicate a yield of only about sixty-two pounds of dressed fiber per acre. On the other hand the letter written by "B," heretofore quoted, speaks of 250 pounds, and I find one Connecticut Yankee reporting yields as high as 350 pounds per acre. It is evident that some exceptional men obtained yields far above the average. Flax was recognized as requiring a fertile, well-drained soil, and yet it is certain that it was produced in quantities sufficient for home needs upon the very poorest of hill farms. Chance references to the weight of straw that would grow upon an acre lead me to believe that the yield of the dressed flax fiber was only about ten per cent of the weight of the flax straw. In the days of our earliest agricultural journalism, dressed flax was one of the few commodities the price of which was regularly quoted in the very brief and primitive market reports of that day. Grown primarily to supply the needs of the home it is evident that there was sometimes a surplus and that this enjoyed a cash market along with wheat, corn, wool, and potash. During a period of ten years I find quotations ranging from seven to thirteen cents a pound. Price varied according to quality and it might be reckoned a very stable commodity.

As has been said, the flax wheel was once found in every well-ordered farm home, and while some of them have been irreverently destroyed and more of them have been "collected," yet there must still remain many thousands of these honorable implements hidden away in the garrets of the old farmhouses of our state. The wheels still survive, but we have seen almost the last of those cunning-fingered women who knew how to use them. High up in the hill country of southern Schoharie County is the onetime prosperous and now almost deserted hamlet of Eminence. In an old farmhouse there lived one Miss Hattie Felter, whose knowledge of the lore of flax spinning still survived to a generation when everybody else had forgotten. Long ago, I used to sit by her wheel and question her, but when she passed she left no successors. Even a generation ago, however, there still remained among us a considerable number of women, some of them not very old, who could deftly spin wool on the wool-spinning wheel. In the evolution of our handicrafts, the spinning of wool persisted for a generation after the spinning of flax had become a lost art. As a matter of fact, there is almost no resemblance between the flax and wool wheels and the mechanics of the two operations are entirely unlike. The flax spinner sat at her work, while her skilled fingers separated out and constantly fed to the foot-turned reel a succession of long filaments drawn from the heavy strand of dressed flax that was thrown over her distaff. All in all it was a fine and gracious art and one in which our foremothers achieved an astonishing proficiency.

Among our old-time farm families, whose ancestral roots run back across the years to early farm occupancy, there are still to be found very many beautiful examples of linen wrought from home-grown flax carried through all the stages of manufacture, spun in the farm kitchen, and woven on the family loom. To me, a fringed linen tablecloth or beautiful bedspread with such a history, if in the possession of the family who made it, seems almost like a patent of nobility.

Weaving is one of the most primitive and ancient of arts. The modern loom has evolved into a wonderfully intricate and complex machine, yet on the huge and clumsy barn-frame loom of a century ago our grandmothers achieved fabrics that bear the stamp of genuine artistry. Of colors they had no great choice. They surely had no coal tar dyes of multifarious shades sold in ten cent packages. From their own fields and woodlands they had certain sources of soft coloring, walnut hulls, butternut bark, and onion skins, but indigo was their main dependence, so most of their patterns were wrought in blue and white.

Less readily than wool did flax adapt itself to machine methods. The manufacture of wool slowly and by degrees passed out of the home to the factory. In our state at least the handling of flax and the weaving of linen always remained a household handicraft and passed away with the coming of the machine age. So it has come at last that a crop, an art, and a handicraft which, less than a century ago, were everywhere are today as forgotten as the quill pen or the flint and steel.

Flax Wheel

CHAPTER XXIII

The Golden Fleece

UNIVERSAL and invaluable as flax might be, it is evident that the pioneer could not be clothed from it alone. Flax furnished a textile that was strong and durable, within limits, almost indestructible, but it did not provide clothing in which to brave the rigors of a New York winter. The very earliest pioneers made much use of furs and skins, raccoon, beaver, fox, bear, and deer, and he who could fling across his bed an imperial bear pelt slept cosily warm regardless of outside temperatures. The first-comers made caps of the thick fur of raccoon, sometimes coats and breeches of deerskin, and even took a lesson from the Indian and went shod in moccasins of buck, but, after all, such dress was possible only for the first thin line of settlement. In the homespun age our state was already pretty well occupied and the spoils of the hunter cut small figure in either the wardrobe or the larder of the farmer. It was necessary almost from the first that a flock of sheep as well as a plot of flax should find a place on every farm.

I think we may safely assume that up until well past the middle of the last century sheep were on every farm. In the year 1845 we had almost six and one-half million in the state and there were two counties, Madison and Otsego, which boasted more than a quarter of a million each, surely a most extraordinary density of ovine population. If we had today in New York State as many sheep per capita as we had then, more than one-half of all the sheep in North America would be within our borders. The 1845 number seems almost unbelievable, but doubtless the sheep of that day were by no means as well bred or developed as are the few that survive today. If we divide the total pounds of wool secured that year by the number of fleeces sheared, the result indicates that the average weight of fleece was only slightly in excess of three pounds. Most of the sheep of that time belonged to an inferior, rather coarse-wooled type, with very little wool on face, legs, or belly. My father, who was an expert shepherd, used to call them "bare-bellys." Nevertheless, there were a few progressive and enthusiastic men who had imported from Europe Merinos of the best type, and the early records of the New York State Agricultural Society demonstrate that some of them were capable of giving very heavy fleeces. Doubtless, there were occasional flocks that were comparable with today's.

During the early years of the past century, there was an enthusiasm for the Merino sheep which eventually became nothing other than a wild craze, perhaps as spectacular and unreasonable as anything that has ever been witnessed in our agriculture. In the volume on manufactures of the federal census for 1860 may be found some accounts of these excesses. Speculative bidding up of values finally carried the price of individual sheep to figures ranging from $500 to $1,500, a boom utterly without reason and the more remarkable because this was a new country without large wealth of any kind. Merino wool shared in the craze. In 1807 it sold at $1.00 per pound and by 1814 the price had climbed to $2.00, $2.50, and even $4.00. Broadcloth was the fashionable fabric, the war had cut off the usual supply from Britain, and American broadcloths sold up to $18.00 per yard. The aftermath of the boom was most disastrous. Within a few years the market was greatly depressed, and the writer adds that thirty years later broadcloth could hardly be sold for $1.00 a yard.

In 1845 there were enough sheep in the state to have given an average of more than thirty to every farm, and there must have been some really large flocks. Many men grew wool for sale, prices during that decade ranging from eighteen to sixty cents per pound according to the year and even more the quality. The lower price was for the coarse, hairy wool of the common un-improved sheep, while the highest was for Saxony. The Saxony breed was a strain of Spanish Merinos that had been bred for a century in Saxony and were considered to be the last word in extremely fine wool. Probably the term was a trade designation for any wool of Merino class and of especially good quality. The fashion in type of wool has always been changeable and it would seem that eighty years ago very fine wool occupied a position of especial favor such as it no longer enjoys. The household handi-crafts of those years required large amounts of wool, and if there was a surplus, it was one of the comparatively few commodities for which there was a dependable cash market. Compared with other staples the price seems good. It is no wonder that at that date the bleating of sheep resounded on every farmstead.

There was a peculiar custom then, possessing all the force of ritual and law, which prescribed that all sheep must be "washed," just previous to shearing, and the date selected was usually the first warm days in May. In my earliest boyhood I frequently witnessed this ceremony. In the early 1870's there were still a great many sheep in our locality, and sometimes on a balmy spring day the road would be full of flocks awaiting their turn to be im-mersed beneath the gushing curtain of water that poured over the spillway of the Wakeman brothers' milldam. There was an arrangement of pens and gates to facilitate the work, and these seem to have constituted a sort of community public utility of which everyone might avail himself.

The flock would be driven into the small pen, and then, one by one, each protesting nanny was dragged beneath the plunging stream and vigorously soused for a minute or two; she was then allowed to rejoin her mate. Long before the job was over the washers were fully as wet as the washed. The spring sunshine might be bright and the air balmy, but the water was still icy cold, so on this one day of the year the strictest total abstainer was not only permitted, but urged, to take a stiff hooker before and after the ceremony, strictly, of course, as a measure of preventive medicine. On the whole, the washing was rather a perfunctory performance, and as a real cleansing measure could have amounted to very little, although it is true that below the washing place the little stream ran sudsy and discolored.

In the days when our state had several million sheep there were many men who proclaimed themselves to be expert shearers, and some of them were worthy of the name. To shear sheep with the old-fashioned shears, quickly, smoothly, and without cutting the animals was a job requiring a skill to be attained only through long experience. Our local expert was one Dan Tinkler, who had learned his trade in England and who went from farm to farm and peeled off the fleeces with practiced skill. In warm weather it was a hot and greasy job. His charge was six cents per head and his earnings were regarded as high wages at that time. The number of sheep some of these old-time workers are alleged to have sheared in a day seems incredible. I believe that a full hundred was the high mark. This is a number three or four times as great as our clumsy operators of today manage, even with the aid of power driven sheep clippers. The old-time shearings, however, were generally of small sheep with no wool to bother on legs, cheeks, or foretop, and through whose coarse, open fleece the shears could be pushed with surprising rapidity. It would have been an entirely different story if they had been asked to deal with wrinkly Merinos or a well-covered, close-wooled Shropshire.

I have not found it possible to gather much information relative to the precise technique of the domestic manufacture of wool. Those elect women who were with us when the homespun age drew to a close and who could have then given such ample and illuminating testimony have since gone the way of all the earth. In the great libraries are shelves of books dealing with modern woolen manufacture, but only here and there can be found a phrase that has any reference to the homespun art. While our earliest periodical agricultural literature has a great deal to say concerning the care and breeding of sheep, it has hardly a word concerning the manufacture of their fleeces. It is a melancholy reflection that a great amount of skill and knowledge concerning one of the most fundamental of household arts has left behind no written word or even tradition.

Wool, when it is shorn, contains a large amount of grease or yolk, a substance which, although it seems greasy, is not a true oil but more akin to a potash soap. In coarse wool there may be as little as twenty per cent of this, while in the case of fine, gummy Merino fleeces it may make up seventy-five per cent or more of the total weight. Modern practice sorts wool from the same fleece into many different grades, and I assume, but without authority, that the old-time spinner also graded wool to some extent, even if less exactly. The next step was to get rid, as far as possible, of all the kinds of grease and dirt. This is now known as scouring and is accomplished by the use of special machinery and various reagents. Our grandmothers carried on by the use of the family washtub, homemade soap, and weak lye. I have been told that where sheep were badly stuck up with burdock, all the care of the worker would not suffice to remove the sharp spines, and a garment made from such wool was a torment if worn next the skin.

Having been washed and dried, the fleece was picked by hand into a soft and fluffy mass and was then ready for carding. Carding, whether done by automatic machines, as today, or by the manual skill of our distaff ancestors, consists essentially in forming the wool into cylindrical rolls, with the wool fiber combed out so as to lie lengthwise of the roll and parallel with each other, in order that they can be drawn out and twisted into yarn by the rapidly revolving spindle. Up until the early years of the past century this work, in America at least, was accomplished entirely by the use of hand cards, implements simple in appearance and almost identical with the familiar wire-studded cards used in grooming cattle. The universal use of hand cards is attested by the statement that as early as 1788 an establishment in Boston made 63,000 of them in a single year. The production of spinning rolls by the use of these cards was an operation calling for no small amount of skill. Its success was dependent, I am told, not only on knack and properly prepared wool but also upon the right temperature. In the exhibit of early implements at the New York State Fair in 1925 there were some of these cards, and two or three women who came out of the onlooking crowds attempted to demonstrate their use but with rather disappointing results. Nevertheless, it was an art in which our grandmothers must have attained a high degree of proficiency.

It is believed that the first carding machine in America was perfected by one Arthur Schofield, an Englishman who, about 1801, established a carding mill at Jamaica, Long Island. At first his charge for carding wool was twelve and a half cents a pound including picking and greasing, greasing being the addition of small amounts of vegetable oil to make the wool work better. Later, other machines were introduced, and the price fell to ten and eventually to eight cents a pound. In the course of time, Schofield began the manufacture of his machines at Pittsfield, Massachusetts, and so it was that Berkshire County, the western county of that state, became the cradle of the woolen industry in America. It is evident that machine carding supplied an acute need, for the introduction of the machines was very rapid. Only a generation after Schofield's first machine they were to be found in almost every township where a stream of water could be made to turn a wooden overshot water wheel.

The carding mill was a community utility, and it was rarely that one was not found within convenient reach in the days when men's usual travel was circumscribed within a radius of a dozen miles. In 1845 there were enumerated in our state 820 carding mills. Ten years later there were found 117 "carding and cloth-dressing establishments." During the next decade the mortality remained high, for almost exactly one-half of these became extinct, and only fifty-nine answered the roll call of 1865. In 1925, after diligent inquiry, we were able to find in the entire state one, possibly two, places where a fleece of wool could be transformed into spinning rolls. The New York State Agricultural Society sent to Harry Cunningham, of Warrensburg, one-hundred pounds of raw wool and, in due course of time, received back some forty-six pounds of spinning rolls, the great shrinkage being accounted for by the loss in scouring and to a small extent in carding. These rolls furnished material for the old-time spinning contest held at the State Fair in 1926, and the skilled spinners pronounced them good rolls.

The wheel for wool spinning was both in construction and in method employed totally different from the flax wheel. A writer in 1860 refers to them respectively as "the great and little spinning wheel." It is evident both from tradition and from the statistics of the flax industry that the domestic manufacture of flax came to an end very soon after the Civil War, but the spinning of wool lingered for a generation thereafter. Indeed, there still remain in our state a considerable number of elect women who in their youth learned to spin wool and who can still gracefully demonstrate that queenly accomplishment, but diligent inquiry established the fact that flax spinning is definitely a lost art. In the garret of the farmhouse of the writer are today two wool and two flax spinning wheels—not collected (I esteem collecting and dealing in antiques as closely akin to grave robbery) but survivals, reminding us that we had once a patriarchal family and a homespun age.

Probably there was never an occupation equal to spinning wool for developing grace and supple poise in a woman. She

Wool Wheel

alternately and rhythmically advanced and retreated with head erect and shoulders thrown far back, while she whirled the wheel with her right hand, and, with her left held aloft, manipulated the thread, and guided it on the spinning spindle. When the spindle was full, it was wound off onto the reel. The circumference of the reel was ordinarily about six and one-half feet. Forty revolutions of the reel (forty threads) was a knot. Ten knots was one skein. Four skeins (approximately two miles of yarn) was a day's work for a busy spinner. To accomplish it she must pace back and forth at least four miles. From her day's work she could knit eight pairs of stockings or ten pairs of mittens or weave two to four yards of cloth. We can hardly conceive the unceasing, almost pitiless industry by virtue of which the pioneer household was clothed.

Cloth for shirts and pantaloons for men, for sheeting, and for table use was woven of pure linen. So, too, bed blankets and dress stuffs for women, and winter garments for men were often of pure wool. There was also very extensive use of linen warp in combination with woolen woof, and long before the final extinction of home weaving, the use of factory-spun cotton warp became very general. The last survival of domestic weaving is the manufacture of rag carpet and silk rugs, a pseudo-homespun industry kept alive by a popular fad for this sort of floor covering.

Much of the cloth woven in the farm homes of the state was used undressed, that is, just as it came from the loom. For more than a hundred and fifty years, however, we have had fulling mills and cloth-dressing shops where the homewoven fabrics were taken for processes not possible in the home. As has been noted, the earliest carding mills were introduced during the very first years of the last century, but in Massachusetts at least a fulling mill is referred to as early as 1764, and by 1790 such mills seem to have been very common. As noted, the state of New York had 820 carding mills in 1845. That same year there were 740 fulling mills, and the two industries were often combined in one building under the same management. Fulling mills and cloth-dressing shops disappeared before the last carding mill came to an end. The household spinning of wool for domestic knitting persisted long after the home weaving of cloth ceased.

I have been unable to find any exact description of the technique of the fuller of that time. Cloth as woven on the domestic loom was virgin wool, but it was somewhat open in texture, and lacked firmness and body. Also it carried a good deal of oil or grease that had been added by the housewife to facilitate carding and spinning. This grease was removed by the use of "fuller's earth," a fine clay which has an affinity for grease that made it more efficient than any soap or washing. The cloth was put into very hot water along with soap and fuller's earth and vigorously soused and beaten with paddles, often by mechanical power. The hot water, soap, and beating caused the wool fibers to felt or mat together. At the same time the cloth shrank in dimensions, but became firmer and heavier, while the soap and fuller's earth made it thoroughly clean. The art of the dyer and fuller were frequently combined.

The cloth dresser carried the work still further. The process of using teasles to raise the nap on cloth was introduced at least as early as 1800, and not long after a machine was devised to shear this nap to a smooth and uniform surface. Doubtless when the housewife received back again the fruit of her loom she found it greatly beautified and improved. I find one single reference to the effect that in Massachusetts, more than a century ago, the price for fulling and dressing domestic cloth was fifty cents per yard—a burdensome charge, considering the low wage and scarcity of money in that time. No wonder that much cloth was used rough as it came from the barn frame loom.

The census enumerators of 1845 made inquiry in every family as to how many yards of cloth had been woven in the home during the preceding year, and the total was more than seven million and ninety thousand yards. About sixty-two per cent of this was woolen and the remainder linen or cotton and linen mixtures. This great amount is a significant commentary on the activity of the textile art in the home of a century ago, or in the childhood of our most aged men and women.

The matronly lore of that time included not only carding, spinning, and weaving, but also a very considerable knowledge of dyeing. Sometimes the yarn was dyed, and at her weaving the craftswoman worked with two different colors. On her primitive loom she managed to secure patterns of very intricate design. Witness the wonderful blue and white linen bedspreads which were the especial field of the textile industry of that day. More commonly the cloth was woven and then dyed to some uniform shade. Indigo was used even in colonial times and it is by common consent the world's most permanent blue dye. Logwood, saffron, cochineal, and annato were among the dyer's limited lists of purchases, and copperas and alum were widely used as mordants. The housewife also knew a considerable list of native roots and barks which gave variety to her hues. From a time-yellowed manuscript book in my possession, I quote these primitive directions for tan color: "One bag tanbark. Boil one half day. Strain. Put in the goods. Boil again a long time. Take out the goods and run through clean lime water. It is better to let the goods remain in the dye three or four days if you can." Butternut bark, hickory nut shucks, and onion skins were on the approved list of domestic dyestuffs.

It is hardly necessary to speak at any length concerning the gentle and placid art of knitting. In early colonial days, two centuries ago, the housewives carded, spun, and dyed yarn from home-grown wool and from it knitted stockings in three colors, the masterpieces of that time. It is safe to say that in the days of which I write all our people wore socks, stockings, wristlets, and comforters, which were the product of this very ancient art. As late as 1831, there was reported only one stocking factory in the whole country, at Newburyport, Massachusetts, and this had "a number of looms operated by females, each of which could produce about twenty stockings per day." Until Civil War times, at least, our farm people knew no other foot covering than the heavy knitted sock or stockings. Even until two generations ago, the farmer who must be abroad in bitter winter weather, felt that if he wore a heavy woolen sock inside a boot that was well greased with a mixture of tallow and beeswax, he enjoyed the acme of possible comfort. My father, born in 1835, often assured me that once a man's boots were well frozen on his feet, he rarely suffered from the cold, provided his socks were thick and his boots waterproofed. No wonder that greasing boots was one of the favorite indoor sports of that time. But it is also true that the chilblain had an importance and standing in the world entirely unknown in these days of felts and overs. Even in my youth I was familiar with them in a way that my son never knew.

Postscript

MY FATHER'S long life was fortunate in that it bridged the transition period between the old and new. He was born on a farm in the days of the scythe, the cradle, the flail, the forty-tooth harrow and the buck saw—and they were not bad days. But he lived to rejoice in the fact that the tractor and automobile and electricity had become a regular part of the farmer's equipment. I do not imagine that I or my son will ever see changes so sweeping and fundamental. Contrary to popular opinion, progress is neither continuous nor eternal.

The foregoing pages have been written with a humble sense of deep admiration mixed with wonder for the courage and resourcefulness of these men and women who in days now almost forgotten fared forth, oftentimes not knowing whither they went, to lay in the wilderness the foundations of an enduring civilization.

Life as they found it seems to us narrow and hard, lacking in privilege or opportunity and filled only with eternal toil. Nonetheless, I am persuaded that to them it seemed a good life. I believe that they, too, found time for laughter and song and love and kisses and withal a certain high hopefulness which we perhaps do not possess.

To bear witness to this, I call one who himself was born when the homespun age was still in flower and who grew up in what was then a pioneer community. Dean Isaac Phillips Roberts, who was

Cornell's great teacher of agriculture, in the evening of a remarkably long life, told his story under the title, *Autobiography of a Farm Boy*. Of his Seneca County farm home he wrote:

I have been a pioneer in three fertile new states in my time, but I have yet to see a country so liberally supplied with the bounties of life, or a people so sturdy, productive, and self-reliant as the inhabitants of the Lake Country of New York, in the second quarter of the Nineteenth Century. My uncle Thomas Burroughs, my father Aaron P. Roberts, and our nearest neighbor Michael Ritter, owned adjoining farms together comprising between four and five hundred acres. There were born to these three heads of families thirty-two children, only one of whom died before reaching the age of thirty and that one lost his life in the battle of Bull Run. These children were all strong and capable; some of them rose to places of modest distinction; all of them were law-abiding and temperate in habits of living and thought, and most of them received all or nearly all of their education in the schools of the district or in the nearby-academies [pp. 25–26].

He fails to add that he, himself, came to be a figure whose long shadow still lies across the field of agricultural education in America. It may add something as an indication of the intellectual activity of some of those early pioneers that every one of the nine children of Aaron Roberts was at some period of his life engaged in teaching.

If the above picture seems idyllic, not to say idealized, may I plead that it was drawn by a very old and much beloved man who saw the scene across a long span of years, and perhaps in retrospect the vista seemed more lovely than it really was. In any case, it supports one of my favorite contentions—that the homespun age nourished many splendid qualities of heart and brain.

Index

Salem (Ore.), 18
Salt industry, 110, 126
Salt pork, 4
Sampson, William, 93
Sap buckets, 107-108, 173
Saratoga County (N.Y.), 244, 247
Sausage, grinding of, 4
Sawmills, 31, 125, 127, 130-132, 138-139
Saws, 28-29
Scandinavia, 250
Schaeffer, John P., 41, 42, 43
Schenectady (N.Y.), 24, 136, 158
Schofield, Arthur, 262
Schoharie County (N.Y.), 3, 29, 45, 56, 59, 60, 62, 115, 116, 120, 141, 159, 184, 191, 226, 237, 254; Child's directory of (1872), 195
Schoharie Creek, 144, 147, 149, 150, 168
Schoharie Republican (1824), quoted, 167-168
Schoharie Valley, 24, 99, 136, 165; settlement of, 36-37
School teachers, qualifications of, 13-14
School year, length of, 14
Scientific American, 150
Scotland, 21, 74, 244
Scythes, 233-235, 244
Seagers, Clayton B., 94, 97
Seneca County (N.Y.), 114, 221, 247, 270
Seneca Indians, 37, 74
Seneca-Oswego Canal, 144
Seneca Turnpike, 151, 153
Settlements, pioneer, progress of, 24
Sewing machines, 192
Shafer, John, 243
Shafer, Merry, 61
Sharon Hill (N.Y.), 215
Shaw, George Bernard, 13
Sheaves, Joseph's dream of, 229
Sheep, 21, 85, 224, 257-261; merino, 258-260; shearing of, described, 260; Shropshire, 260; washing of, described, 259-260
Sheffield Scientific School (Yale), 226
Sheraton, Thomas, 113
Sherman, Foster and Company (merchants), 224
Shields, Greek, agricultural symbols on, 229

Shingles, shaving of, 141 ff; tools for, 142
Shipbuilding, 104-105
Ships, names of, 105
Shoemakers, 192-199, 201-202; tools of, 199-200
Shoepegs, wooden, 156, 195-196, 200
Shoes, 3; ankle length, 196; making of, 191-192; repairing of, 193; sewing of, described, 199-200
Shoestrings, 196
Shops, boot-and-shoe, 194
Sickles, 225
Silk, production of, 180-182
Simms, Jeptha, 165
"Sith," 225-226
"Slashing," described, 30
Sleeplessness, cure for, 81
Sleighs, making of, 116-121
Smelters, iron, 10
Snaths, described, 234
Snyder Family, 177
Soap, making of, 3
Soils, 222
"Solitary Reaper" (William Wordsworth), quoted, 30
Souse, 5
South America, 85, 86, 190
South Carolina, 180, 246
Sowing, 223-224
Soy beans, 80
Spain, 85
Spinning, 251, 255, 264
Spinning contest (1926), 263
Spinning wheels, 3; wool, described, 263
Springfield (Mass.), 154
Squash, 87, 92
Stahl, Austin, 229
Stanton, James, 142
"State Bridge," 149
Staten Island (N.Y.), 39
Steamboats, 25, 158
Stockings, 3, 267
Stone Age, 103
Stones, gathering of, described, 235
Stores, country, commodities of, 9-10
Straw: flax, 254; rye, 242
Stumps, nature and destruction of, 32
Succotash, 88
Sugar, 174; maple, 108, 110
Sullivan County (N.Y.), 39-40
Sullivan's Expedition, 88